Robot manipulators are becoming increasingly important in research and industry, and an understanding of statics and kinematics is essential to solving problems of robotics.

This book provides a thorough introduction to statics and first-order instantaneous kinematics with applications to robotics. The emphasis is on serial and parallel planar manipulators and mechanisms.

The text differs from others in that it is based solely upon the concepts of classical geometry. It is the first to describe how to introduce linear springs into the connectors of parallel manipulators and to provide a proper geometric method for controlling the force and motion of the hands (end effectors) or tools of serial robot manipulators performing constrained motion tasks.

Both students and practicing engineers will find this book easy to follow, with a clearly written text and abundant illustrations, as well as exercises and real-world projects to work on.

Professor Joseph Duffy is Director of the Center for Intelligent Machines and Robotics at the University of Florida.

Statics and kinematics with applications to robotics

Statics and kinematics

with applications

to robotics

JOSEPH DUFFY

University of Florida

CAMBRIDGE
UNIVERSITY PRESS

CAMBRIDGE UNIVERSITY PRESS
Cambridge, New York, Melbourne, Madrid, Cape Town, Singapore, São Paulo

Cambridge University Press
The Edinburgh Building, Cambridge CB2 2RU, UK

Published in the United States of America by Cambridge University Press, New York

www.cambridge.org
Information on this title: www.cambridge.org/9780521482134

First published 1996
This digitally printed first paperback version 2006

A catalogue record for this publication is available from the British Library

Library of Congress Cataloguing in Publication data
Duffy, Joseph, 1937–
Statics and kinematics with applications to robotics / Joseph
Duffy.
p. cm.
Includes index.
ISBN 0-521-48213-5 (hardcover)
1. Manipulators (Mechanism)   2. Statics.   3. Machinery, Kinematics
of.   4. Robots – Motion.   I. Title.
TJ211.D84   1996   95-4913
670.42′72 – dc20   CIP

ISBN-13 978-0-521-48213-4 hardback
ISBN-10 0-521-48213-5 hardback

ISBN-13 978-0-521-03398-5 paperback
ISBN-10 0-521-03398-5 paperback

To my wife, Anne

Contents

vii

Preface

This text is devoted to the statics of rigid laminas on a plane and to the first-order instantaneous kinematics (velocities) of rigid laminas moving over a plane. Higher-order instantaneous kinematic problems, which involve the study of accelerations (second-order properties) and jerk (third-order properties) are not considered.

This text is influenced by the book *Elementary Mathematics from an Advanced Standpoint: Geometry*, written by the famous German geometer Felix Klein. It was published in German in 1908 and the third edition was translated into English and published in New York by the Macmillan Company in 1939. The book was part of a course of lectures given to German High School Teachers at Göttingen in 1908. Klein was admonishing the teachers for not teaching geometry correctly, and he was essentially providing a proper foundation for its instruction.

The present text stems from the undergraduate course "The Geometry of Robot Manipulators," taught in the Mechanical Engineering Department at the University of Florida. This course is based on Klein's development of the geometry of points and lines in the plane and upon his elegant development of mechanics: *"A directed line-segment represents a force applied to a rigid body. A free plane-segment, represented by a parallelogram of definite contour sense, and the couple given by two opposite sides of the parallelogram, with arrows directed opposite to that sense, are geometrically equivalent configurations, i.e., they have equal components with reference to every coordinate system."*

It is the author's opinion that a student's understanding of statics and instantaneous kinematics is enhanced immeasurably by learning that quantities
such as force and turning moment have a geometrical meaning, and that they
are equivalent to a directed line segment and a directed area, respectively.
This facilitates understanding the nature of invariance and explains how the
behavior of such quantities is influenced by the group of Euclidean transformations.

It is sad that the great developments in geometry of the last century and
its application in mechanics have, for the most part, been forgotten or ignored by many researchers in the field of robotics. Perhaps the most blatant
exclusion is that no reference is made by modern researchers in robotics to
the theory of constrained motion, which was completed by Sir Robert Stawell
Ball in his monumental *Treatise on the Theory of Screws* (1900, Cambridge
University Press). Because of this omission, a large number of erroneous articles have been presented at national and international robotics conferences
and in the technical journals of learned societies. These articles aim to model
constrained-motion tasks, grasping operations, and the calibration of robots.
Ball correctly modeled the constrained motion of a rigid body by permissi-
ble, instantaneous rigid-body motions together with constraint forces/couples
in combinations, which he defined well over a century ago as, respectively,
twists of freedom and wrenches of constraint.

Many researchers either are unaware of Ball's work and/or choose to ig-
nore it. They introduce fallacious concepts, such as "the motions a constrained
body cannot have," which they bravely define as "natural constraints." It con-
cerns me to see these errors enshrined in a growing number of textbooks on
robotics for engineering students. Such texts subject students to meaningless
expressions in which dimensions and units are mixed, for example, (Linear
Velocity)^2 + (Angular Velocity)^2, and present answers to problems in which
the values of physical quantities change with the choice of reference point
(i.e., the answers are not invariant with a translation of origin). Students are
also subjected to least-squares minimization problems with objective functions such as $(d^2 + \theta^2)$, where d and θ, respectively, have units of length and
radians. These incorrect concepts are devoid of geometric meaning, and they
can have no physical meaning either. As we advance toward the 21st century, the rigorous study of statics and instantaneous kinematics applied to the

field of robotics is apparently receding and being regarded as archaic. The author hopes that this short text will help to reverse this trend.

The author wishes to thank Dr. E. J. F. Primrose (formerly Mathematics Department, University of Leicester), Dr. J. M. Rico-Martínez (Instituto Tecnológico de Celaya), and Dr. M. G. Mohamed (El-Mina University) for their valuable comments and suggestions on the text. The author also wishes to thank Dr. M. Griffis, Mr. I. Baiges, Mr. R. Hines, and Mr. S. A. Kelkar (University of Florida) for their contributions to the text. The author is indebted to Ms. Cindy Townsend for typing the manuscript, and to Ms. Florence Padgett, Editor, Physical Sciences, Cambridge University Press, for her advice and guidance in the preparation of the text.

J. Duffy
Gainesville, Florida
October 1995

Introduction

Statics and first-order instantaneous kinematics are intimately related. They are completely analogous or, more specifically, they are dual concepts as, for example, a line and a point are dual in the projective plane. The meet of two lines is always a point in the projective plane, on which parallel lines are said to meet at a point at infinity. This proposition can be re-stated for two points by making the appropriate grammatical changes in order for it to make sense. The dual statement is simply that the join of two points is a line. It is always possible to formulate (prove) a proposition (theorem) for one dual element and to simply state a corresponding proposition (theorem) for the corresponding dual element. Another example for lines and points is:

Two distinct lines which have a common point determine a planar pencil of lines through that point.

Any two distinct points determine a range of points on the line which joins the two points.

In statics a line segment represents the "linear" concepts of force, whereas in instantaneous kinematics a line segment represents the "circular" concept of instantaneous rotation.

This book evolved from lecture notes for a first course in robot manipulators, an undergraduate course in mechanical engineering at the University of Florida, which was first offered in the fall of 1989. The major objective of the course is to provide students with a proper understanding of statics and first-order instantaneous kinematics. The prerequisites are that students

1

have passed entry level courses in statics and dynamics and that they have a working knowledge of a computer language.

This text is primarily (Chapters 2–5) devoted to a study of planar statics and first-order instantaneous kinematics with applications to manipulators. However, to perform a static and instantaneous kinematic analysis of a manipulator it is necessary to know the geometry of the manipulators. That is, we need to know the relative angles between consecutive pairs of links connected by turning joints and the relative linear displacements between pairs of consecutive links connected by sliding joints. An analysis of the geometry of planar manipulators is given in Chapter 1.

The elements of mechanisms and manipulators are also introduced in Chapter 1. This is followed by determining the so-called forward and reverse position analysis of planar, serial two- and three-jointed manipulators. At the outset, the student is required to draw a planar serial manipulator, and to identify and label joint variables with manipulator dimensions. Equations are then obtained by parallel projection. A forward analysis is relatively straightforward. A unique position for a point in the end effector together with the end-effector orientation can be computed for a specified set of joint variables.

The same equations are used to perform a reverse analysis; the coordinates of a point in the end effector are specified and usually two configurations (elbow up and elbow down) are computed for a two-jointed serial manipulator. Analogously, the coordinates of a point in the end effector together with the end-effector orientation for a serial three-jointed manipulator are specified, and a pair of elbow up and elbow down configurations are computed.

Chapter 1 concludes with a reverse analysis of a parallel manipulator with the general geometry for which a rigid lamina (called a moving platform) is connected to the ground via three revolute-slider-revolute kinematic chains, typically called connectors. Here the reverse solution is relatively straightforward. The specification of the location (position and orientation) of the moving platform yields a unique solution for the three connector lengths. However, specifying the three connector lengths leads to multiple forward solutions for the location of the moving platform. For the most general case, there may be as many as six locations, which is complicated. Finally, an example of a parallel manipulator with a special geometry that yields four locations of the moving platform is given.

Chapter 2 develops the planar statics of rigid laminas based upon the firm foundation of geometry. It follows closely the development of F. Klein, and forces and moments are shown to be geometrically equivalent to directed line segments and areas, respectively. A force can be considered a scalar multiple of a line, and homogeneous line coordinates are synonymous with force coordinates.

A study of a system of forces is essentially a study of a system of lines. The resultant force that acts upon the platform of a parallel device is a linear combination of the system of forces generated in the connectors of the platform (forward static analysis). Conversely, an external force that acts upon the platform will produce a component resultant in each platform connector (reverse static analysis).

Chapter 3 develops the planar, first-order, instantaneous kinematics of laminas. As in Chapter 2, this chapter follows closely the development of F. Klein. An instantaneous rotation of a rigid lamina is essentially a scalar multiple of a line, which is the rotation axis.

A study of the relative instantaneous motion of a number of serially connected laminas is really a study of the lines which define the various axes of rotation. The instantaneous motion of the end effector of a serial planar manipulator is a linear combination of the rotation speeds of revolute joints or the linear displacement speeds of sliding joints in the chain (forward kinematic analysis). Conversely, when the instantaneous motion of the end effector of a serial manipulator is specified, the joint motions required to produce the required end-effector motion can be calculated (reverse kinematic analysis).

It should become clear to the student that not only are statics and kinematics analogous, or dual, but the statics of a parallel manipulator is completely analogous, or dual, with the instantaneous kinematics of a serial manipulator. That is, the resultant force that acts upon the end effector (the resultant instant motion of the end effector) of a parallel manipulator (of a serial manipulator) is a linear combination of the connector forces (instant joint motions). Furthermore, the forward and reverse analyses of the statics and instantaneous kinematics of parallel and serial manipulators are shown to be dual.

Chapter 4 establishes the complete duality of statics and kinematics, and shows that the forward and reverse statics analyses of a serial device and the forward and reverse instantaneous kinematics of a parallel device are also dual. Again, as in the previous chapters, this chapter is based on a firm foundation of geometry by introducing such invariant quantities as instantaneous power and instant work, which, for the instantaneous rotation of a lamina produced by a force in the plane of motion, are scalar multiples of the mutual moment of a pair of lines, the rotor axis, and the line of action of the force.

The concepts of the mutual moment of a pair of lines, instant power, and instant work date back well over a century, as do the conditions for the vanishing of instant power (or work) discovered by Sir Robert Stawell Ball. The invariant condition for the vanishing of instantaneous power was defined by Ball as *reciprocity*, and he based his theory of constraints and freedoms of partially constrained rigid bodies on this definition. These are the same concepts that have largely been ignored by many modern researchers in robotics. A new theory, called the modern hybrid control theory, for the control of force and motion has surfaced. This theory is based on a condition that is not invariant with a choice of reference point. It yields equations with mixed units, has no geometrical meaning whatsoever, and is therefore nonsense.

Chapter 4 concludes with an infinitesimal displacement analysis of a parallel manipulator together with the derivative of a line rotating in a plane about an axis normal to the plane of rotation. These expressions are used in Chapter 5 to compute a relatively new stiffness mapping, which relates a small change of force acting upon a platform of a parallel manipulator, initially in static equilibrium, to the instantaneous rotation of the platform produced by the change of force. The spring matrix is a function of the spring constants of linear springs introduced into the legs together with the preload and the geometry. In 1965 F. M. Dimentberg derived a spring matrix for a parallel manipulator which remains close to its unloaded configuration.

It is well known that a small change in force δf can be produced by a small displacement δx of a linear spring, and $\delta f = k\delta x$, where k is the spring constant. This means that a small change in force could be controlled by the small displacement of a spring. Chapter 5 extends this concept to simultaneously control the force and the displacement of a point connected to the ground via a pair of linear springs. Following this, it is extended to simultaneously control the force and motion of a rigid lamina, the platform of a parallel manipulator connected to the ground via three compliant revolute-slider-revolute connectors. A linear spring is inserted in each slider joint.

A variety of numerical examples are given throughout the text. It is advantageous for students to have access to computer graphic facilities so they can animate manipulators and display their results.

1

A displacement analysis of planar robot manipulators

1.1   Mechanisms and manipulators

A rigid lamina is free to move on a fixed reference plane, the page. The lamina is unbounded and covers the entire reference plane. However, for ease of visualization an arbitrary contour is drawn on the moving lamina, and the moving lamina is usually identified with the area enclosed by the contour.

The motion of the lamina measured relative to the fixed plane can be determined by first specifying some reference point O in the fixed plane. In Figure 1.1, O is chosen outside the contour (it could equally be chosen on or inside the contour boundary). Next, some reference point Q on the moving lamina is chosen; its position relative to O can be measured, for instance, by the coordinates (x_Q, y_Q). Finally, a reference line is drawn on the moving lamina, and its orientation γ is measured relative to some reference line in the fixed plane, for example, the X axis. Changes $\{\Delta x_Q, \Delta y_Q; \Delta \gamma\}$ in the three parameters $\{x_Q, y_Q; \gamma\}$* can be used to quantify the motion of the moving lamina. Since each parameter can be varied independently, the moving plane is said to possess three freedoms or three degrees of freedom (3 d.o.f.) with respect to the fixed plane.

It is important to recognize that there are many representations of the location and, hence, the motion of a moving lamina. One is free to select the

* The units for γ and (x_Q, y_Q) are, respectively, radians or degrees and length (inches, feet, or meters, etc.). The semicolon is introduced to indicate this difference in units.

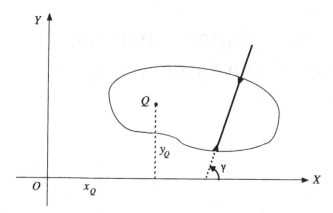

Figure 1.1 The location of a lamina in a fixed reference plane.

two reference points O and Q together with the orientation of the XY coordinate system.

The moving lamina may not possess three d.o.f. It may be constrained, for example, to rotate about an axis Z, normal to the reference plane, by a revolute R kinematic pair, hinge, or turning joint (see Fig. 1.2). Clearly, all points on the moving lamina move on concentric circles centered on the axis of rotation. Here, it is convenient to draw reference lines through point O on the axis of rotation and to measure the motion by changes in the single parameter θ. A revolute pair permits one freedom ($f = 1$) of the moving lamina relative to the fixed plane.

Mechanisms and robot manipulators are illustrated by closed and open polygons that use such reference lines. These skeletal forms are essentially geometrical models which can be labeled conveniently with the joint variables and the link lengths.

The moving lamina may be constrained to pure translational motion by a prismatic P kinematic pair (sliding joint). A prismatic pair has no special line because all points on the moving lamina translate in a specific direction. It is, however, common practice to designate the center line of the joint as a reference line and to measure the displacement of some point Q on the moving lamina relative to some point O in the fixed plane which lies on this reference line.

An assemblage of links and joints such as those illustrated in Figure 1.2 was defined by Reuleaux (1876) as a kinematic chain that may be open or closed.

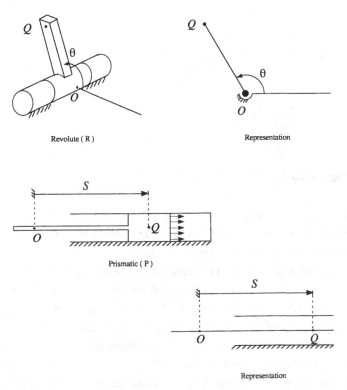

Figure 1.2 Revolute and prismatic kinematic pairs.

It can be a single open or closed loop, or it can be a combination of open and closed loops. Reuleaux stated: "In itself a kinematic chain does not postulate any definite absolute displacement.* One must hold fast or fix in position one link of the chain relatively to the portion of surrounding space assumed to be stationary. The relative displacements of links then become absolute. A closed kinematic chain of which one link is made stationary is called a mechanism."

The link that is held fixed is called the frame or the frame of reference. A change in the selection of a reference frame is known as *kinematic inversion.*

---

* The term "displacement" has been substituted for the term "motion" used in Reuleaux's original text. The motion of a rigid lamina relative to a reference frame implies not only displacement but velocity, acceleration, and so on (see Hunt 1978).

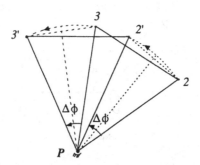

Figure 1.3 Finite rotation pole.

The relative displacement between any pair of links is independent of the choice of reference frame. On the contrary, the absolute displacement of links is dependent on the choice of reference frame.

A link 23 (see Fig. 1.3) can be displaced to a second position 2'3' by rotating it about a point P, commonly called the pole, fixed in the reference frame, the page (the ground), which is the point of intersection of the perpendicular bisectors of 22' and 33'. Point P can be modeled by a revolute pair that connects to the ground the moving lamina which contains link 23.

This same displacement can be achieved by connecting points 2 and 3, respectively, to any pair of points 1 and 4 which lie on the perpendicular bisectors S and T of 22' and 33' (see Fig. 1.4). Revolute joints are now located at the four vertices of the quadrilateral and, in this way, a 4R mechanism is formed. Link 41 is stationary and link 23 (the coupler) undergoes an anticlockwise rotation $\Delta\phi$ about P when an actuator drives either link 12 or 43 in an anticlockwise direction.

Assume that link 23 is held fixed, and it is thus considered the frame of reference. The same relative displacement of the four links of the mechanism can be obtained by rotating link 41 clockwise about P, as shown in Figure 1.5. The quadrilaterals 412'3' and 4'1'23 are congruent. However, the absolute displacements of the links are different.

Assume that the links of the 3R open chain 1234 are given successive anticlockwise angular displacements of 30, 40, and 90 degrees beginning with link 12 connected to the ground via the revolute pair (see Fig. 1.6). Then suppose that link 34 is connected to the ground via a revolute joint, and the links are given successive clockwise displacements of 90, 40, and 30 degrees, be-

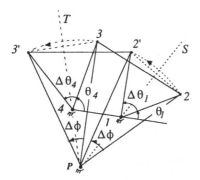

Figure 1.4 A 4*R* planar mechanism.

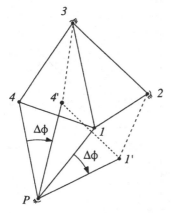

Figure 1.5 A kinematic inversion of the 4*R* planar mechanism.

ginning with link 34. It is clear from the figure that the relative displacements of the links for the two cases are the same, and the chains 12'3'4' and 1"2"3"4 are congruent. However, the absolute displacements of the links are different.

The closed-loop kinematic chain 4123 with link 41 as the frame of reference is usually called a mechanism. There is one driving link 12 (the crank), and the movement of the mechanism is repeated through each revolution of the crank.

The open kinematic chain 1234 with link 12 connected to the ground via a revolute is an example of a serial robot manipulator where an end effector or some form of gripping device is attached at the free end. Each revolute joint is actuated so that the relative displacements of link 12 to the ground,

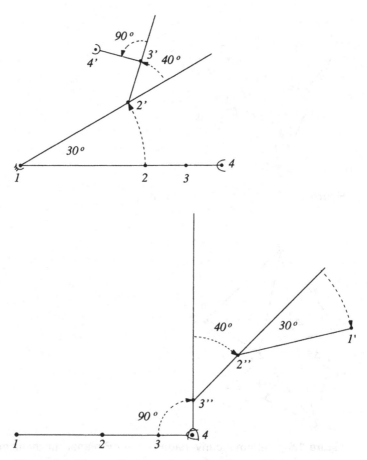

Figure 1.6 Kinematic inversions of a planar 3*R* open chain.

23 to 12, and 34 to 23 are controlled independently. This device is capable of performing a multitude of tasks in contrast to the previous mechanism, which is capable of only cyclic movement.

Robot manipulators are not necessarily serial. Figure 1.7 illustrates two versions of a planar manipulator with a movable triangle 123 connected to the ground (a fixed triangle 1′, 2′, 3′) by three *RPR* chains in parallel. This device is called an in-parallel robot manipulator, and it is capable of performing a multitude of tasks by actuating each of the three prismatic pairs or, alternatively, each of the three grounded revolute pairs.

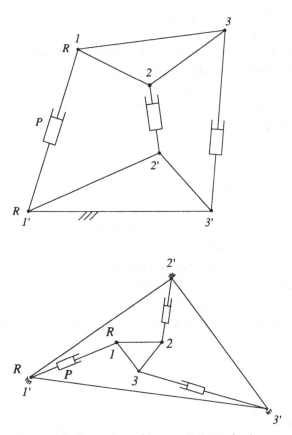

Figure 1.7 Examples of in-parallel manipulators.

1.2   The mobility of planar mechanisms and manipulators

The closed planar quadrilateral (see Fig. 1.4), the open serial chain (see Fig. 1.6), and the parallel assemblages (see Fig. 1.7) have been given mobility by introducing revolute and prismatic pairs between the rigid bodies that form the assemblages. We may now derive the mobility criterion for an assemblage of rigid links and kinematic pairs.

We have established that the rigid lamina (or link), which is free to move on the XY plane (see Fig. 1.1), possesses three degrees of freedom. Clearly, if there were n such unconnected links lying on the XY plane the assemblage would possess $3n$ independent degrees of freedom. Also, if one link were fixed in the XY plane, the assemblage would possess $3(n-1)$ freedoms. Assume that any pair of links is connected by at most one kinematic pair with freedom f. Then the relative freedoms between the pair of links is reduced by $(3-f)$. Therefore, the degrees of freedom of the n links connected by j kinematic pairs are reduced to $\sum_{i=1}^{j}(3-f_i)$, where i denotes the ith kinematic pair. The overall number of freedoms of the assemblage are commonly defined as its mobility M, and

$$M = 3(n-1) - \sum_{i=1}^{j}(3-f_i). \tag{1.1}$$

Equation 1.1 is derived without any consideration of the special geometry of the assemblage; it cannot be applied without modification to assemblages which, for example, contain two or more prismatic joints that line up, because this duplicates the sliding motion of a single prismatic joint. When the arrangement of pairs is general these results are obtained:

(i) For a closed single-loop assemblage the number of pairs equals the number of links, $j = n$, and (1.1) reduces to

$$M = \sum_{i=1}^{n} f_i - 3. \tag{1.2}$$

For all $f_i = 1$ and for $n = 3, 4, 5 \ldots$, $M = 0, 1, 2 \ldots$, i.e., a triangle, a quadrilateral, and a pentagon . . . possess, respectively, zero (a structure), one, and two freedoms. When the quadrilateral has one link fixed to the ground (see Fig. 1.4) it requires a single input drive to control the motion, which is repetitive and cyclic. This assemblage is usually called a four-bar mechanism. If one link of a hinged pentagon were fixed to the ground it would require two input drives to control the motion.

(ii) For an open serial chain the first pair is connected to ground while the end link is free to move in the XY plane. For this case, $n = 1$ (ground) $+ n_c$ (the number of moving links in the chain). Here $n_c = j$ and thus $n = 1 + j$. Equation 1.1 reduces to

$$M = \sum_{i=1}^{j} f_i. \tag{1.3}$$

Therefore, serial chains with one end connected to the ground, with all $f_i = 1$ and $j(= n_c) = 1, 2, 3, 4 \ldots$, possess mobility one, two, three, four. . . . Such assemblages are employed as manipulators. An end effector is attached at the free end. An end effector is capable of a variety of motions and of performing a multitude of tasks as opposed to the closed-loop, four-bar mechanism, which is only capable of a single cyclic motion. However, note that a single input drive is required to actuate a four-bar mechanism, whereas a drive is required to actuate each joint in a serial chain.

Finally, it is interesting to determine the mobility M of the parallel assemblage illustrated by Figure 1.7. Clearly, there are three *RPR* chains that connect the base to a movable lamina and therefore $j = 9$. Furthermore, there are two links in each of the three chains that connect the base to the movable lamina and hence $n = 8$. The substitution of these values in (1.1) yields $M = 3(8 - 1) - 9 \times 2 = 21 - 18 = 3$. It follows that three input drives must be employed to control the motion of the movable lamina. Either electric motors can be used to actuate the three base revolute pairs or pneumatic (hydraulic) actuators installed to actuate the three sliding joints in each *RPR* chain.

1.3 Displacement analysis of single degree of freedom (d.o.f.) planar manipulators

The assemblages illustrated in Figure 1.8 can be considered one d.o.f. planar manipulators, where the moving lamina is a link connected to the ground by a revolute pair actuated by a motor or a prismatic pair actuated by a hydraulic or pneumatic cylinder. At the free end, there is a gripper or hand on which a reference point Q is selected (Q could also designate a point on a tool held by the gripper). The link length is known, and the single variables θ and S are, respectively, controlled and the corresponding gripper position is Q ($a \cos \theta$, $a \sin \theta$) and $Q(S, a)$, where a is the distance of Q from the center line of the joint. The gripper itself may be actuated, but such motions are independent of the gross motion of the robot manipulators and are of no concern here.

Note that the skeletal form of link a, which connects the tool point Q to the prismatic joint, is drawn perpendicular to the slider displacement S, and its length is measured from the center line of the joint, Q. It could be drawn at any angle α. Clearly, this cannot affect the relative sliding motion, and the

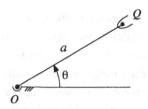

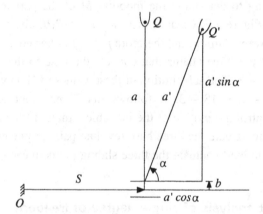

Figure 1.8 Revolute and prismatic manipulators.

coordinates for the new tool point Q' are simply $(S + a' \cos \alpha, a' \sin \alpha + b)$. From here on, links that connect prismatic joints will be drawn perpendicular to the direction of sliding.

1.4   Displacement analysis of two-link serial manipulators

It is convenient to describe serial manipulators by their sequence of kinematic pairs, beginning with the first or grounded kinematic pair. Figure 1.9 illustrates the four, two-link, two d.o.f. *RR*, *RP*, *PR*, and *PP* manipulators. Here, both kinematic pairs in each serial chain are actuated. The relative angular displacement between a pair of links a_{ij} and a_{jk} ($ij = 01$ (ground), 12; $jk = 12, 23 \ldots$) is denoted by θ_j and variable slider displacements are labeled S_j, $j = 1, 2. \ldots$. The reader will recall (see Fig. 1.8) that the center line of each prismatic pair is designated as a reference line. All link lengths are measured from these center lines. For example, the link length a_{23} of the

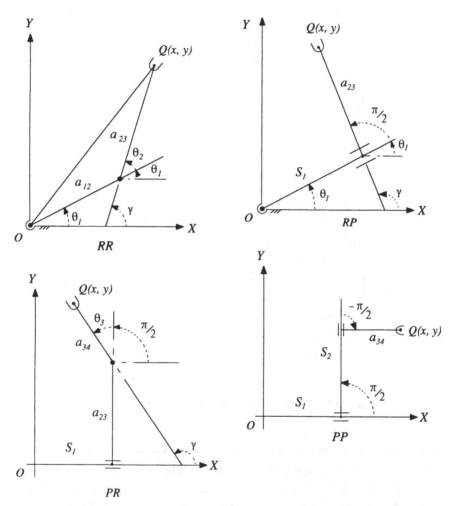

Figure 1.9 Planar two-link manipulators.

RP and *PR* manipulators is measured as the perpendicular distance from the center lines labeled S_1.

The analysis of two- and three-link manipulators involves the solution of two trigonometrical equations. The first equation is

$$d \cos \theta = f, \tag{1.4}$$

where the coefficients d and f are known quantities.

Provided that $|f/d| \leq 1$, when $f/d > 0$, a calculator will display a value $\theta^{(a)}$ in the first quadrant ($\pi/2 > \theta^{(a)} > 0$) of the unit circle (see Fig. 1.10). Since

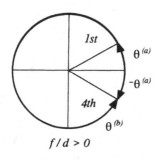

f/d > 0

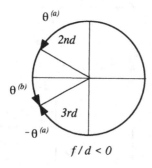

f/d < 0

π

Figure 1.10 Solutions of cos θ = f/d.

the cosine is an even function, a second solution is given by $-\theta^{(a)}$ or $\theta_b = (2\pi - \theta^{(a)})$, which is in the fourth quadrant. When $f/d < 0$, a calculator will display a value $\theta^{(a)}$ in the second quadrant ($\pi > \theta^{(a)} > \pi/2$) of the unit circle, and a second solution is $-\theta^{(a)}$ or $\theta_b = (2\pi - \theta^{(a)})$, which is in the third quadrant. The pair of values $\theta^{(a,b)}$ denotes two distinct manipulator configurations for a specified end-effector location. It is important to identify and to keep track of these configurations in path planning the end effector.

The second equation is

$$e \sin \theta = g, \tag{1.5}$$

where the coefficients g and e are known quantities.

Provided that $|g/e| \leq 1$, when $g/e > 0$ a calculator will display a value $\theta^{(a)}$ in the first quadrant of the unit circle (see Fig. 1.11). A second solution is given by $(\pi - \theta^{(a)})$, which is in the second quadrant, and is denoted by $\theta^{(b)}$.

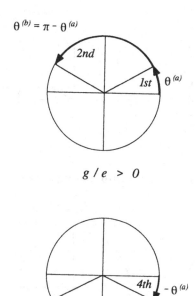

$$\theta^{(b)} = \pi - \theta^{(a)}$$

2nd

1st $\theta^{(a)}$

$$g/e > 0$$

4th $-\theta^{(a)}$

3rd

$$\theta^{(b)} = \theta^{(a)} - \pi$$

$$g/e < 0$$

Figure 1.11 Solutions of $\sin \theta = g/e$.

When $e/g < 0$ a calculator will display a value $-\theta^{(a)}$ in the fourth quadrant. A second solution is given by $-(\pi + (-\theta^{(a)}))$, which is in the third quadrant and is denoted by $\theta^{(b)}$.

Finally, we need to solve the following pair of simultaneous trigonometrical equations:

$$A \cos \theta - B \sin \theta = E, \tag{1.6}$$

and

$$B \cos \theta + A \sin \theta = F, \tag{1.7}$$

where A, B, E, and F are known coefficients. Subtracting $B \times$ (1.6) from $A \times$ (1.7) and then adding $A \times$ (1.6) to B $\times$ (1.7) gives, respectively,

$$\sin \theta = \frac{(AF - BE)}{(A^2 + B^2)}, \tag{1.8}$$

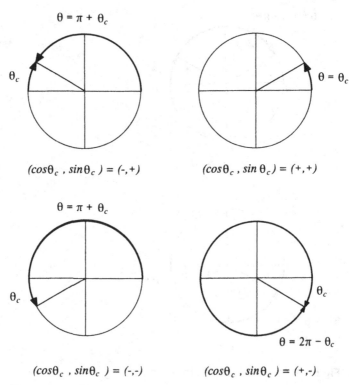

$\theta = \pi + \theta_c$

θ_c

$(cos\theta_c , sin\theta_c) = (-,+)$

$\theta = \theta_c$

$(cos\theta_c , sin\theta_c) = (+,+)$

$\theta = \pi + \theta_c$

θ_c

$(cos\theta_c , sin\theta_c) = (-,-)$

θ_c

$\theta = 2\pi - \theta_c$

$(cos\theta_c , sin\theta_c) = (+,-)$

Figure 1.12 Computation of θ.

and

$$\cos \theta = \frac{(AE + BF)}{(A^2 + B^2)} , \qquad (1.9)$$

provided that either A or B is different from zero.

The pair of values ($\sin \theta$, $\cos \theta$) yields a single value for θ. This is easy to compute from the value $\theta_c = \tan^{-1} \{(AF - BE)/(AE + BF)\}$, which is obtained by dividing (1.8) by (1.9). A calculator will display either a positive or negative value for θ_c. It remains to determine in which quadrant the angle θ lies. A calculator will display a positive value for θ_c when θ lies in the first and third quadrants (see Fig. 1.12), and a negative value for θ_c when θ lies in the second and fourth quadrants. The quadrant can be identified by the ordered pair of the signs of ($\cos \theta_c$, $\sin \theta_c$), as illustrated by the unit circles in Figure 1.12.

1.4.1 Analysis of the *RR* manipulator

Throughout, the coordinates of point Q will be denoted by (x, y), which by parallel projection (see Fig. 1.9) on the X and Y axes are

$$x = a_{12}c_1 + a_{23}\,c_{1+2}, \tag{1.10}$$

$$y = a_{12}s_1 + a_{23}\,s_{1+2}. \tag{1.11}$$

Also,

$$\gamma = \theta_1 + \theta_2. \tag{1.12}$$

The abbreviations $c_1 = \cos\theta_1$, $s_1 = \sin\theta_1$, $c_{1+2} = \cos(\theta_1 + \theta_2)$, and $s_{1+2} = \sin(\theta_1 + \theta_2)$ have been introduced.

When the joint displacements θ_1, θ_2 are known, (1.10) and (1.11) determine a unique location (position and orientation) for the gripper $\{x, y; \gamma\}$. This is called the *forward analysis.*. The *reverse analysis* is more difficult. We need to specify a position of a tool point Q in the end effector and to compute the joint displacements θ_1 and θ_2. This is accomplished by adding the sums of the squares of (1.10) and (1.11), which yields

$$(x^2 + y^2) = a_{12}^2 + a_{23}^2 + 2a_{12}a_{23}\,(c_{1+2}c_1 + s_{1+2}s_1). \tag{1.13}$$

Then introduce the identity $c_{1+2}c_1 + s_{1+2}s_1 = c_{1+2-1} = c_2$ and regroup terms to give

$$d \cos\theta_2 = f, \tag{1.14}$$

where

$$d = 2a_{12}a_{23},$$

$$f = x^2 + y^2 - a_{12}^2 - a_{23}^2. \tag{1.15}$$

Equation 1.14 yields two solutions, θ_2 and $-\theta_2$, or equivalently $\theta_2^{(a,b)}$ (see the solution of 1.14), which define a pair of manipulator configurations (see Fig. 1.13). Corresponding pairs of values $\theta_1^{(a,b)}$ can be computed by solving (1.10) and (1.11) for s_1 and c_1. This is accomplished by expanding the right sides of (1.10) and (1.11), and re-grouping terms, which yields

$$(a_{12} + a_{23}c_2)c_1 - (a_{23}s_2)s_1 = x, \tag{1.16}$$

$$(a_{23}s_2)c_1 + (a_{12} + a_{23}c_2)s_1 = y. \tag{1.17}$$

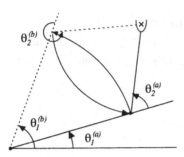

Figure 1.13 Two *RR* manipulator configurations.

These equations are the same form as (1.6) and (1.7) with $A = a_{12} + a_{23}c_2$, $B = a_{23}s_2$, $E = x$, and $F = y$. A pair of values $\theta_1^{(a,b)}$ which corresponds to $\theta_2^{(a,b)}$ can be computed. By using (1.12), the sets $(\theta_1^{(a)}, \theta_2^{(a)})$ and $(\theta_1^{(b)}, \theta_2^{(b)})$ determine uniquely the orientations $\gamma^{(a)}$ and $\gamma^{(b)}$.

1.4.2 Analysis of the *RP* manipulator

By parallel projection on the X and Y axes (see Fig. 1.9), the coordinates of Q are

$$x = S_1c_1 + a_{23}\cos\left(\frac{\pi}{2} + \theta_1\right) = S_1c_1 - a_{23}s_1, \tag{1.18}$$

$$y = S_1s_1 + a_{23}\sin\left(\frac{\pi}{2} + \theta_1\right) = S_1s_1 + a_{23}c_1. \tag{1.19}$$

$$\gamma = \theta_1 + \pi/2. \tag{1.20}$$

When the joint displacements θ_1; S_1 are known, (1.18)–(1.20) determine a unique location for the gripper $\{x, y; \gamma\}$. The reverse analysis is performed by adding the sums of the squares of (1.18) and (1.19) and solving for S_1^2, which yields

$$S_1^2 = x^2 + y^2 - a_{23}^2. \tag{1.21}$$

Equation 1.21 gives a pair of values, $S_1 = \pm(x^2 + y^2 - a_{23}^2)^{1/2}$, which will be denoted by $S_1^{(a,b)}$. Equations 1.18 and 1.19 are the same form as (1.6) and

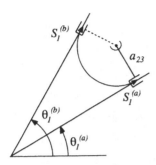

Figure 1.14 Two *RP* manipulator configurations.

(1.7), with $A = S_1^{(a,b)}$, $B = a_{23}$, $E = x$, and $F = y$. A pair of values $\theta_1^{(a,b)}$, which corresponds to $S_1^{(a,b)}$, can thus be computed. Finally, a pair of values $\gamma^{(a,b)}$ can be computed from (1.20) (see Fig. 1.14).

1.4.3 Analysis of the *PR* manipulator

By parallel projection on the X and Y axes (see Fig. 1.9), the coordinates of Q are

$$x = S_1 + a_{34} \cos(\pi/2 + \theta_3) = S_1 - a_{34}s_3, \tag{1.22}$$

$$y = a_{23} + a_{34} \sin(\pi/2 + \theta_3) = a_{23} + a_{34}c_3, \tag{1.23}$$

$$\gamma = \theta_3 + \pi/2. \tag{1.24}$$

When the joint displacements $\{S_1; \theta_3\}$ are known, (1.22)–(1.24) determine a unique location for the gripper $\{x, y; \gamma\}$.

A reverse analysis is performed by solving (1.23) for θ_3, which yields two solutions, $+\theta_3$ and $-\theta_3$, or equivalently, $\theta_3^{(a,b)}$ (see the solution of (1.4)), which defines a pair of manipulator configurations (see Fig. 1.15). Corresponding values for the slider displacement $S_1^{(a,b)}$ can be computed from (1.22) and a corresponding pair of values $\gamma^{(a,b)}$ can be computed from (1.24).

1.4.4 Analysis of the *PP* manipulator

By parallel projection on the X and Y axes (see Fig. 1.9), the coordinates of Q are

$$x = S_1 + a_{34}, \tag{1.25}$$

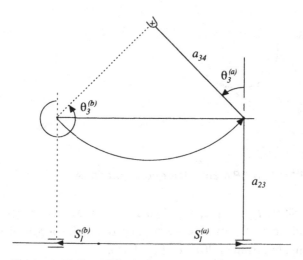

Figure 1.15 Two *PR* manipulator configurations.

and

$$y = S_2. \tag{1.26}$$

Furthermore,

$$\gamma = 0. \tag{1.27}$$

The forward solution is given by (1.25) and (1.26), whereas the reverse solution is given by $S_1 = x - a_{34}$ and $S_2 = y$. There is a unique solution for a specified value of the coordinates of Q.

   Complete Exercises 1.1 and 1.2 as far as possible without using the text by employing the following procedure.

1.   Draw the skeletal form of the manipulator to be analyzed using the sequence of joints specified, beginning with the first grounded joint.

2.   Label the skeletal form with the manipulator dimensions and variables.

3.   Employ parallel projection, write expressions for the coordinates (x, y) of a tool in the end effector, and obtain an expression for the orientation of the end effector measured relative to the X axis.

4.   Perform a reverse analysis by solving the expressions for (x, y) for one joint variable.

5.   Compute a pair of sets of joint variables which define the two distinct closures for the end-effector location by employing, where necessary, the solutions of the trigonometrical equations 1.4–1.7.

EXERCISE 1.1

We want to position the tool points Q of planar RR, RP, and PR manipulators (see Figs. 1.13–1.15).

Manipulator type	Dimensions*	Coordinates of Q
RR	$a_{12} = 1.5, a_{23} = 0.5$	(0.317, 1.549)
RP	$a_{23} = 1$	(0.134, −2.232)
PR	$a_{23} = 2, a_{34} = 1.5$	(−2.207, 1.293)

Perform a reverse analysis for each manipulator, compute values for γ, and draw to scale in inches the two configurations for each manipulator. Verify your results by performing a forward analysis.

1.5   Analysis of 3R and 2R-P three-link serial manipulators

The notation for two-link serial manipulators (see Section 1.4) is now adapted to the three-link RRR, RRP, RPR, and PRR manipulators (see Fig. 1.16).** Some readers may ask why a PPP manipulator is not included in this list. The answer is that a nonrotating PPP assemblage can only produce pure translational motion of the end effector, which can be obtained by actuating a PP manipulator (see Fig. 1.17). A third P joint is not linearly independent of the first two P joints.

---

* The dimensions and coordinates can be chosen in any consistent units of length.

** Section 1.4 examined cases where the end effector only has two degrees of freedom. In other words, we are not able to freely choose all three parameters $\{x, y; \gamma\}$. With the addition of a third independent joint (a P or R joint), it is possible to control the three freedoms of the end effector, specifically by controlling the three joints simultaneously. This section develops the analytics necessary to do this.

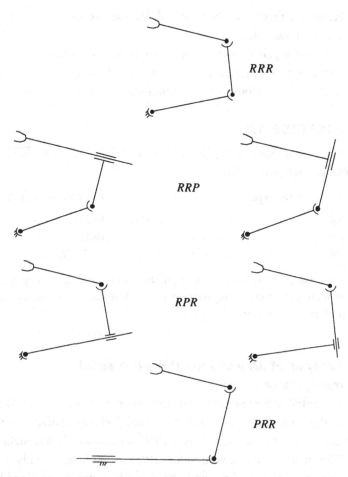

Figure 1.16 3*R* and 2*R-P* manipulators.

1.5.1 Analysis of the *RRR* manipulator

By parallel projection on the X and Y axes, the coordinates of point Q are (see Fig. 1.18)

$$x = a_{12} c_1 + a_{23} c_{1+2} + a_{34} c_{1+2+3}, \tag{1.28}$$

$$y = a_{12} s_1 + a_{23} s_{1+2} + a_{34} s_{1+2+3}. \tag{1.29}$$

Furthermore, the orientation angle is

$$\gamma = \theta_1 + \theta_2 + \theta_3. \tag{1.30}$$

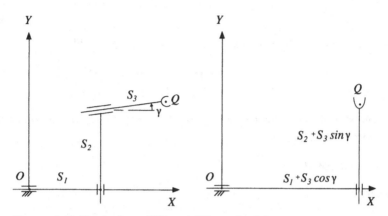

Figure 1.17 Equivalent *PPP* and *PP* manipulators.

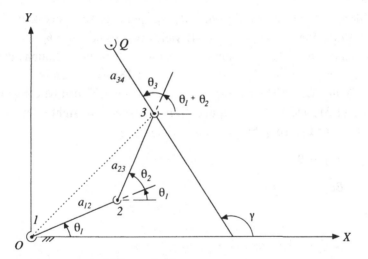

Figure 1.18 An *RRR* manipulator.

When the joint displacements θ_1, θ_2, and θ_3 are known, (1.28)–(1.30) determine a unique location for the gripper $\{x, y; \gamma\}$. A reverse analysis for which $\{x, y; \gamma\}$ is specified is accomplished by substituting (1.30) into (1.28) and (1.29) and re-arranging in the form

$$x - a_{34} \cos \gamma = a_{12}c_1 + a_{23}c_{1+2},$$ (1.31)

$$y - a_{34} \sin \gamma = a_{12}s_1 + a_{23}s_{1+2}. \tag{1.32}$$

Add the sums of the squares of (1.31) and (1.32) to yield

$$(x - a_{34} \cos \gamma)^2 + (y - a_{34} \sin \gamma)^2$$
$$= a_{12}^2 + a_{23}^2 + 2a_{12}a_{23}(c_{1+2}c_1 + s_{1+2}s_1). \tag{1.33}$$

Then, we introduce the identity $c_{1+2}c_1 + s_{1+2}s_1 = c_{1+2-1} = c_2$ and regroup the terms, which gives

$$d \cos \theta_2 = f, \tag{1.34}$$

where

$$d = 2a_{12}a_{23},$$

$$f = (x - a_{34} \cos \gamma)^2 + (y - a_{34} \sin \gamma)^2 - a_{12}^2 - a_{23}^2. \tag{1.35}$$

Equation 1.34 can be obtained directly by applying the cosine law to triangle 123 (see Fig. 1.18). Also, (1.34) yields two solutions $+\theta_2$ and $-\theta_2$, or equivalently, $\theta_2^{(a,b)}$ (see the solution of (1.4)). These two solutions define a pair of manipulator configurations (see Fig. 1.18) which can be constructed using Figure 1.13.* The corresponding values for $\theta_1^{(a,b)}$ can be computed by solving (1.31) and (1.32) for s_1 and c_1. If we expand the right sides of (1.31) and (1.32) and regroup by terms, this yields

$$Ac_1 - Bs_1 = E, \tag{1.36}$$

$$Bc_1 + As_1 = F, \tag{1.37}$$

where

$$A = a_{12} + a_{23}c_2, \quad E = x - a_{34} \cos \gamma,$$

$$B = a_{23}s_2, \quad\quad\quad F = y - a_{34} \sin \gamma. \tag{1.38}$$

These equations have the same form as (1.6) and (1.7). A pair of values $\theta_1^{(a,b)}$

---

* Note that for the 3R manipulator all three coordinates $\{x, y; \gamma\}$ are computed and therefore the two configurations are determined with link a_{34} in a fixed position. It follows that the two configurations can be obtained using the 2R manipulator constructions illustrated in Fig. 1.13. Similarly, other three-link manipulator configurations can be related to corresponding two-link manipulators.

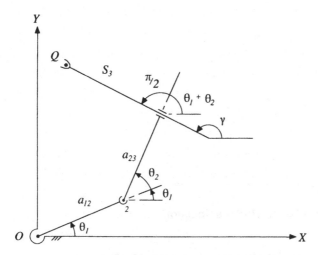

Figure 1.19 An *RRP* manipulator.

can be computed, which corresponds to $\theta_2^{(a,b)}$. Equation 1.30 can be used to compute $\theta_3^{(a,b)}$, which corresponds to sets $\theta_1^{(a)}$, $\theta_2^{(a)}$, and $\theta_1^{(b)}$, $\theta_2^{(b)}$.

1.5.2 Analysis of the *RRP* manipulator

By parallel projection on the X and Y axes, the coordinates of point Q are (see Fig. 1.19)

$$x = a_{12}c_1 + a_{23}\,c_{1+2} + S_3 \cos(\pi/2 + \theta_1 + \theta_2), \tag{1.39}$$

$$y = a_{12}s_1 + a_{23}\,s_{1+2} + S_3 \sin(\pi/2 + \theta_1 + \theta_2). \tag{1.40}$$

The orientation angle is

$$\gamma = \pi/2 + \theta_1 + \theta_2. \tag{1.41}$$

When the joint displacements θ_1, θ_2, and S_3 are known, (1.39)–(1.40) determine a unique location for the gripper $\{x, y: \gamma\}$.

A reverse analysis for which $Q\{x, y; \gamma\}$ is specified is now performed. From (1.41), $-s_{1+2} = \cos\gamma$ and $c_{1+2} = \sin\gamma$. Substitution into (1.39) and (1.40) yields

$$x - a_{23} \sin\gamma = a_{12}c_1 + S_3 \cos\gamma, \tag{1.42}$$

$$y + a_{23} \cos\gamma = a_{12}s_1 + S_3 \sin\gamma. \tag{1.43}$$

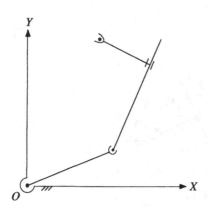

Figure 1.20 An *RRP* manipulator.

Subtract cos γ times (1.43) from sin γ times (1.42) to get

$$a_{12}\sin(\gamma - \theta_1) = x\sin\gamma - y\cos\gamma - a_{23}. \tag{1.44}$$

From (1.41), sin $(\gamma - \theta_1) = \sin(\theta_2 + \pi/2) = c_2$. Therefore, (1.44) can be expressed in the form

$$d\,c_2 = f, \tag{1.45}$$

where $d = a_{12}$ and $f = x\sin\gamma - y\cos\gamma - a_{23}$. This yields two solutions, $+\theta_2$ and $-\theta_2$, or equivalently, $\theta_2^{(a,b)}$ (see the solution of (1.4)). The corresponding pair of values $\theta_1^{(a,b)}$ is computed from (1.41), and a corresponding pair of values $S_3^{(a,b)}$ is computed from either (1.42) or (1.43), and

$$S_3^{(a,b)} = \frac{x - a_{23}\sin\gamma - a_{12}c_1^{(a,b)}}{\cos\gamma} = \frac{y + a_{23}\cos\gamma - a_{12}s_1^{(a,b)}}{\sin\gamma}. \tag{1.46}$$

Figure 1.20 illustrates another form of the *RRP* manipulator. It is left to the reader to perform a forward and reverse analysis.

1.5.3 Analysis of the *RPR* manipulator

By parallel projection on the X and Y axes, the coordinates of point Q are (see Fig. 1.21)

$$x = S_1c_1 - a_{23}s_1 + a_{34}\cos(\pi/2 + \theta_1 + \theta_3), \tag{1.47}$$

$$y = S_1s_1 + a_{23}c_1 + a_{34}\sin(\pi/2 + \theta_1 + \theta_3). \tag{1.48}$$

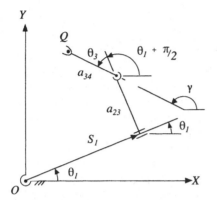

Figure 1.21 An *RPR* manipulator.

The orientation angle is

$$\gamma = \pi/2 + \theta_1 + \theta_3. \tag{1.49}$$

When the joint displacements θ_1, S_1, and θ_3 are known, (1.47)–(1.49) determine a unique location for the gripper $\{x, y; \gamma\}$.

A reverse analysis is performed using (1.47) and (1.48), which can be rearranged as

$$S_1 c_1 - a_{23} s_1 = x - a_{34} \cos \gamma \tag{1.50}$$

$$S_1 s_1 + a_{23} c_1 = y - a_{34} \sin \gamma. \tag{1.51}$$

Adding the sums of the squares of (1.50) and (1.51) yields

$$S_1^2 = (x - a_{34} \cos \gamma)^2 + (y - a_{34} \sin \gamma)^2 - a_{23}^2. \tag{1.52}$$

Equation 1.52 gives a pair of values, $S_1 = \pm\{(x - a_{34} \cos \gamma)^2 + (y - a_{34} \sin \gamma)^2 - a_{23}^2\}^{1/2}$, which will be denoted by $S_1^{(a,b)}$. Equations 1.50 and 1.51 have the same form as (1.6) and (1.7) with $A = S_1^{(a,b)}$ $B = a_{23}$, $E = x - a_{34} \cos \gamma$, and $F = y - a_{34} \sin \gamma$. Thus, a pair of values $\theta_1^{(a,b)}$, which corresponds to $S_1^{(a,b)}$, can be computed. Finally, a pair of values $\theta_3^{(a,b)}$ can be computed from (1.49).

Figure 1.22 illustrates another form of the *RPR* manipulator. It is left to the reader to perform a forward and reverse analysis.

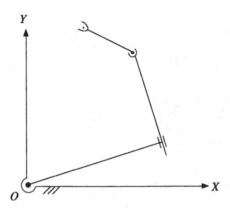

Figure 1.22 Another form of the *RPR* manipulator.

1.5.4 Analysis of the *PRR* manipulator

By parallel projection on the X and Y axes, the coordinates of point Q are (see Fig. 1.23)

$$x = S_1 + a_{23}c_2 + a_{34}c_{2+3}, \tag{1.53}$$

$$y = a_{23}s_2 + a_{34}s_{2+3}. \tag{1.54}$$

The orientation angle is

$$\gamma = \theta_2 + \theta_3. \tag{1.55}$$

When the joint displacements are known, (1.53)–(1.55) determine a unique location for the gripper $\{x, y; \gamma\}$.

A reverse analysis is performed by substituting (1.55) into (1.54), which gives

$$a_{23}s_2 = y - a_{34} \sin \gamma. \tag{1.56}$$

This is in the same form as (1.5), with

$$e = a_{23}, \quad g = y - a_{34} \sin \gamma. \tag{1.57}$$

Hence, the solution yields pairs of values of θ_2, which we denote by $\theta_2^{(a,b)}$

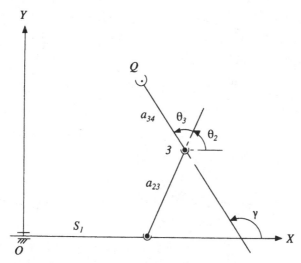

Figure 1.23 A *PRR* manipulator.

(see the solution of (1.5)). These define two distinct configurations, which are easy to construct (using Fig. 1.15). The corresponding values $\theta_3^{(a,b)}$ can be computed from (1.55), and the corresponding values for S_1 can be computed from (1.53):

$$S_1^{(a,b)} = x - a_{23}c_2^{(a,b)} - a_{34} \cos \gamma. \qquad (1.58)$$

EXERCISE 1.2

1. Locate (position tool point Q and orientate (γ) link a_{34}) the end effectors of the *RRR*, *RRP*, *RPR*, and *PRR* manipulators.

Manipulator type	Dimensions (ins.)	Coordinates of Q (ins.)	Orientation γ deg.
RRR	$a_{12} = 3, a_{23} = 2,$ $a_{34} = 0.5$	$(-0.84, 4.02)$	105
RRP	$a_{12} = 3, a_{23} = 2$	$(-3.5, 2.10)$	270
RPR	$a_{23} = 2, a_{34} = 1$	$(-2.5, -3.08)$	145
PRR	$a_{23} = 2, a_{34} = 1$	$(-1.75, 2.7)$	105

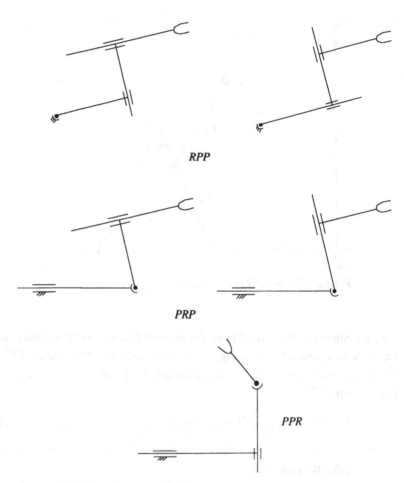

Figure 1.24 The 2*P-R* manipulators.

Perform a reverse analysis for each manipulator and draw the two configurations. Verify your results by performing forward analysis.

1.6   Analysis of 2P-R three-link serial manipulators

The forward and reverse analyses of the three-link *RPP*, *PRP*, and *PPR* manipulators (see Fig. 1.24) are extremely simple. In any one case, there is only a single manipulator configuration for each gripper location $\{x, y; \gamma\}$, because there is a one-to-one relationship between the single revolute joint displacement and γ. It is left to the reader to perform these analyses.

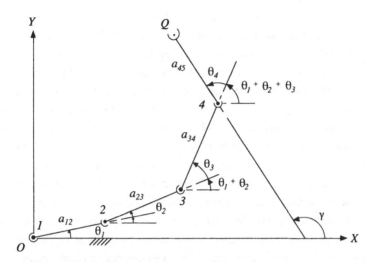

Figure 1.25 A 4R manipulator.

1.7 Analysis of redundant-serial manipulators with four or more kinematic pairs

The example in Figure 1.25 illustrates a planar $4R$ redundant manipulator. By parallel projection on the X and Y axes, the coordinates of point Q are

$$x = a_{12}c_1 + a_{23}c_{1+2} + a_{34}c_{1+2+3} + a_{45}c_{1+2+3+4}, \qquad (1.59)$$

$$y = a_{12}s_1 + a_{23}s_{1+2} + a_{34}s_{1+2+3} + a_{45}s_{1+2+3+4}. \qquad (1.60)$$

The orientation angle is

$$\gamma = \theta_1 + \theta_2 + \theta_3 + \theta_4. \qquad (1.61)$$

When the joint displacements are known, (1.59)–(1.61) yield a unique location for the gripper $\{x, y; \gamma\}$.

However, there is an infinite number of solutions for the reverse analysis. The substitution of (1.61) in (1.59) and (1.60) yields

$$x = a_{12}c_1 + c_{23}c_{1+2} + a_{34}s_{1+2+3} + a_{45} \cos \gamma, \qquad (1.62)$$

and

$$y = a_{12}s_1 + a_{23}s_{1+2} + a_{34}s_{1+2+3} + a_{45} \sin \gamma. \qquad (1.63)$$

It is not possible, in general, to obtain a unique solution for this pair of equations when the gripper location $Q(x, y; \gamma)$ is specified. The $4R$ manipulator possesses an additional freedom, or more than the three freedoms that are sufficient to locate the gripper. Such manipulators are redundant.

The analysis of redundant manipulators has been the subject of much research. Briefly, such manipulators should permit sophisticated articulation and motion planning where the extra freedoms can be used to avoid obstacles. Proposed solutions to the problem usually include optimization in one form or another. Optimization formulations must be geometrically meaningful and thus be invariant under the different selections of reference points, such as O and Q, the orientations of coordinate systems, and they must provide the same results using different systems of units (feet or meters, for example). An interesting problem occurs when a gripper is performing a cyclic motion, i.e., it is repeating a sequence of locations throughout the cycle. It is desirable for any optimization formulation to determine manipulator configurations throughout a cycle which are reproduced in subsequent cycles. A meaningful solution to redundant serial manipulators that accomplishes this is given by Chung, Griffis, and Duffy (1994).

1.8   Displacement analysis of in-parallel manipulators

1.8.1   Description of in-parallel manipulators
The most general form of an in-parallel planar manipulator with three parallel, serially connected RPR chains joining a fixed base to a movable lamina is shown in Figure 1.26. The term in-parallel means that each serial chain has the same structure and one joint is actuated in each chain (for instance, the prismatic pair).

In contrast to serial devices it is relatively simple to perform a reverse analysis, i.e., it is simple to compute a unique set of connector lengths $1 - 1' = \ell_1$, $2 - 2' = \ell_2$, and $3 - 3' = \ell_3$ for a specified location of the moving lamina. However, the forward analysis of the general in-parallel manipulator is algebraically complicated, and we need to compute the multiple locations of the moving lamina for a specified set of connector lengths ℓ_1, ℓ_2, and ℓ_3. The solution can be expressed in the form of a sixth-degree polynomial, which means that there can be up to a maximum of six different real assembly configurations of the mechanism for a specified set of connector lengths. Such an analysis is beyond the scope of this book. However, a method for deriving a sixth-

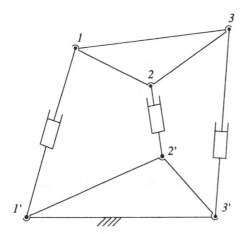

Figure 1.26 A general in-parallel manipulator.

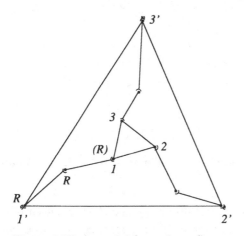

Figure 1.27 A model for a three-finger grasp.

degree polynomial is given in Li and Matthew (1987). This mechanism can be used to model a three-fingered grasp (see Fig. 1.27). The central movable triangle is formed by the contact points of three, two d.o.f. *RR* fingers. Each contact point is also modeled by a revolute (*R*) kinematic pair.

Two in-parallel manipulators with special geometries are illustrated in Figures 1.28 and 1.29.

The gross motions or work spaces of in-parallel manipulators are smaller

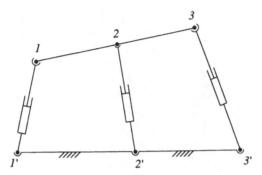

Figure 1.28 An in-parallel manipulator with special geometry (first example).

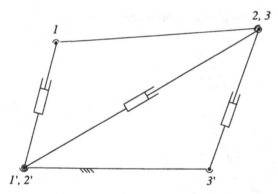

Figure 1.29 An in-parallel manipulator with special geometry (second example).

than those for serial manipulators. However, parallel devices can sustain much greater forces than serial devices. Generally, the advantages of in-parallel manipulators are that they are more accurate in positioning and orientating workpieces than serial manipulators, they possess a high payload/weight ratio, and they are easily adaptable to force and position control.

1.8.2 A reverse analysis for the general in-parallel manipulator

As we mentioned earlier, it is relatively simple to perform a reverse analysis for this device. The location $\{x, y; \gamma\}$ of the movable lamina is specified, where (x, y) are the coordinates of a point Q on the movable platform and γ is the orientation of the platform, measured relative to the X axis.

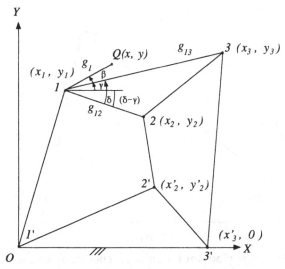

Figure 1.30 A line diagram for a general in-parallel manipulator.

Here γ is measured using side 13. Since point Q is fixed on the moving plat-form, then g_1 and β (see Fig. 1.30) are known as well as the parameters δ, g_{12}, and g_{13} which specify the triangular movable platform. What follows is one of many possible procedures for determining the connector lengths ℓ_1, ℓ_2, and ℓ_3. It is immediately apparent from Figure 1.30 that

$$x_1 = x - g_1 \cos (\beta + \gamma),$$

$$y_1 = y - g_1 \sin (\beta + \gamma).$$

Also,

$$x_2 = x_1 + g_{12} \cos (\delta - \gamma),$$

$$y_2 = y_1 - g_{12} \sin (\delta - \gamma),$$

and

$$x_3 = x_1 + g_{13} \cos \gamma,$$

$$y_3 = y_1 + g_{13} \sin \gamma. \tag{1.64}$$

Finally, $\ell_1 = (x_1^2 + y_1^2)^{1/2}$, $\ell_2 = \{(x_2 - x_2')^2 + (y_2 - y_2')^2\}^{1/2}$, and $\ell_3 = \{(x_3 - x_3')^2 + y_3^2\}$.

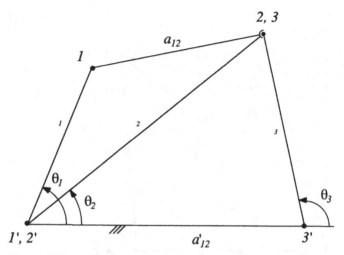

Figure 1.31 A line diagram for an in-parallel manipulator.

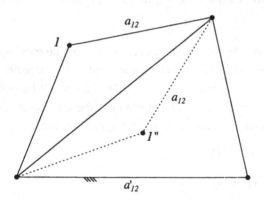

Figure 1.32 Two assembly configurations above the base a'_{12}.

1.8.3 A forward analysis for an in-parallel manipulator with the simplest geometry

As we mentioned before, it is complicated to perform a forward analysis for the general in-parallel manipulator. However, the analysis is relatively simple for this manipulator with special geometry (see Fig. 1.31). For a specified set of connector lengths ℓ_1, ℓ_2, ℓ_3 and sides a_{12}, a'_{12}, two distinct assembly configurations above the base can be obtained. Point $1''$ is a reflection

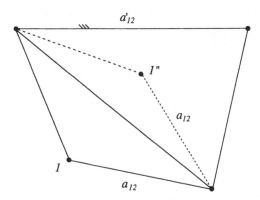

Figure 1.33 Two assembly configurations below the base a'_{12}.

of point 1 through the connector ℓ_2 (see Fig. 1.32). There are two additional configurations (see Fig. 1.33) that are reflections through the base.

The four assemblages in Figures 1.32 and 1.33 can be computed using the following pair of cosine laws (see Fig. 1.31):

$$\cos(\theta_1 - \theta_2) = (\ell_1^2 + \ell_2^2 - a_{12}^2)/2\ell_1\ell_2,$$
$$\cos\theta_2 = (\ell_2^2 + a'^2_{12} - \ell_3^2)/2\ell_2 a'_{12}. \tag{1.65}$$

All four assembly configurations can be drawn if the angles θ_1 and θ_2 are known.

——— **EXERCISE 1.3** ————————————————————

Solve (1.65) and obtain all four assembly configurations of the simplest in-parallel manipulator with $a'_{12} = 3$, $a_{12} = 1.35$, $\ell_1 = 1.95$, $\ell_2 = 3$, and $\ell_3 = 2.2$ (ins.). Verify these results by construction using only a ruler and a compass.

2

Planar statics

2.1 The coordinates of a line in the *XY* plane

Assume that the page is the *XY* plane and the *Z* axis is pointing upward out of the page through *O*. Most students are familiar with using point coordinates, for example, how to designate points 1 and 2 by the coordinates (x_1, y_1) and (x_2, y_2) (see Fig. 2.1). The coordinates of a point are essentially an ordered pair of real numbers. This may be something new for some students: we will determine the coordinates for designating a line.

The two vectors $\mathbf{r}_1$ and $\mathbf{r}_2$ from *O* to points 1 and 2 with coordinates (x_1, y_1) and (x_2, y_2) determine the directed line segment 12 with vector $\mathbf{S}$ given by

$$\mathbf{S} = \mathbf{r}_2 - \mathbf{r}_1. \tag{2.1}$$

The projections of $\mathbf{S}$ onto the *X* and *Y* axes are, respectively,

$$L = x_2 - x_1 = |\mathbf{S}| \cos \theta \quad \text{and} \quad M = y_2 - y_1 = |\mathbf{S}| \sin \theta, \tag{2.2}$$

where

$$|\mathbf{S}| = (L^2 + M^2)^{1/2}. \tag{2.3}$$

$L/|\mathbf{S}|$ and $M/|\mathbf{S}|$ are the direction cosines of the line segment. When $|\mathbf{S}| = 1$, the directed line segment has unit length $L^2 + M^2 = 1$, and then $L = \cos \theta$ and $M = \sin \theta$.

40

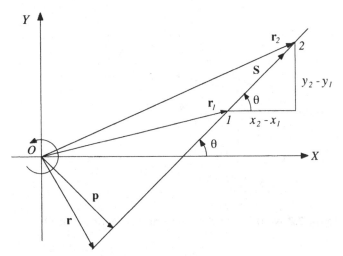

Figure 2.1 A line in the *XY* plane.

The moment of the directed line segment 12 about the origin is given by the vector product $\mathbf{r} \times \mathbf{S}$, where $\mathbf{r}$ is *any* vector drawn from O to any point on the straight line joining points 1 and 2, and the vector $\mathbf{S}$ can be located anywhere on the line.

Now from Figure 2.2,

$$\mathbf{r}_1 \times \mathbf{S} = |\mathbf{r}_1| \cdot |\mathbf{S}| \sin \phi_1 \mathbf{k} = (|\mathbf{r}_1| \sin \phi_1) |\mathbf{S}|\mathbf{k} = |\mathbf{p}| \, |\mathbf{S}|\mathbf{k},$$

$$\mathbf{r}_2 \times \mathbf{S} = |\mathbf{r}_2| \cdot |\mathbf{S}| \sin \phi_2 \mathbf{k} = (|\mathbf{r}_2| \sin \phi_2) |\mathbf{S}|\mathbf{k} = |\mathbf{p}| \, |\mathbf{S}|\mathbf{k},$$

$$\mathbf{r} \times \mathbf{S} = |\mathbf{r}| \cdot |\mathbf{S}| \sin \phi \mathbf{k} = (|\mathbf{r}| \sin \phi) |\mathbf{S}|\mathbf{k} = |\mathbf{p}| \, |\mathbf{S}|\mathbf{k},$$

and

$$\mathbf{p} \times \mathbf{S} = |\mathbf{p}| \cdot |\mathbf{S}| \sin \pi/2 \, \mathbf{k} = |\mathbf{S}|\mathbf{k} = |\mathbf{p}| \, |\mathbf{S}|\mathbf{k},$$

where $\mathbf{k}$ is a unit vector parallel to the Z axis.

Therefore, the moment can be expressed in the alternative forms

$$\mathbf{r} \times \mathbf{S} = \mathbf{r}_1 \times \mathbf{S} = \mathbf{r}_2 \times \mathbf{S}. \tag{2.4}$$

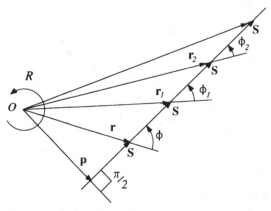

Figure 2.2 Moment of directed line segment about *O*.

The left and intermediate sides of (2.4) can be expressed in the determinant forms

$$\begin{vmatrix} i & j & k \\ x & y & 0 \\ L & M & 0 \end{vmatrix} = \begin{vmatrix} i & j & k \\ x_1 & y_1 & 0 \\ L & M & 0 \end{vmatrix} = \begin{vmatrix} i & j & k \\ x_1 & y_1 & 0 \\ x_2 - x_1 & y_2 - y_1 & 0 \end{vmatrix}$$

$$= \begin{vmatrix} i & j & k \\ x_1 & y_1 & 0 \\ x_2 & y_2 & 0 \end{vmatrix}, \tag{2.5}$$

where **j** and **i** are unit vectors parallel to the *X* and *Y* axes, and *x* and *y* are the components of *r*.

If we equate the left and right sides of (2.5), the **i** and **j** components vanish identically, while the **k** component yields the equation for the line, which can be written in the form

$$Ly - Mx + R = 0, \tag{2.6}$$

where

$$R = \begin{vmatrix} x_1 & y_1 \\ x_2 & y_2 \end{vmatrix}. \tag{2.7}$$

It is left to the reader to deduce (2.6) using the left and right sides of (2.4).

In addition, expanding the right side of (2.5) and equating to the left side of (2.4) yields

$$\mathbf{r} \times \mathbf{S} = R\mathbf{k}. \tag{2.8}$$

The moment vector of the line segment about the origin is parallel to the Z axis and is thus perpendicular to the XY plane. For convenience, the vector $R\mathbf{k}$ in Figure 2.2 is associated with the Z axis. However, subsection 2.4.1 demonstrates that $R\mathbf{k}$ is a *free vector* which can be associated with *any line* drawn parallel to that Z axis.*

The three numbers L, M, and R were first established by Plücker, so they are called the Plücker line coordinates. They are homogeneous because substituting λL, λM, and λR, where λ is a nonzero scalar into (2.6) yields the same line. *However, their units are not consistent. L* and M have dimensions of $(\text{length})^1$, while R has dimensions of area, $(\text{length})^2$. Because of this lack of consistency in dimensions, the coordinates are represented by the ordered triple of real numbers $\{L, M; R\}$ with the semicolon separating R from L and M.

It is important to recognize that if the quantities L, M, and R are known, there remains one degree of freedom in the location of the directed line segment 12, because it takes four magnitudes (x_1, y_1) and (x_2, y_2) to fix the points 1 and 2. The same triple $\{L, M; R\}$ is obtained if and only if the vector $(\mathbf{r}_2 - \mathbf{r}_1)$ is free to move on a definite straight line, and this line is determined when only the ratios $L: M: R$ are known. It should now be clear that the only changes of the position vectors $\mathbf{r}_1$ and $\mathbf{r}_2$ that leave $\{L, M; R\}$ unchanged are translations of the line segment along its line which preserve its length and sense. *Such a line segment, which is determined by the ordered triple of real numbers $\{L, M; R\}$, is called a line bound vector.*

The two coordinates (L, M) by themselves determine a *free vector*, because they are unaltered by a parallel translation of the line segment outside the line. The ratios $L: M: R$ (see also (2.6)), which are equivalent to *two quantities*, determine the *straight line* and *not* the *length of the segment upon it.*

---

* The symbol $\circlearrowright$ is intended to convey the sense of the turning moment R about $\mathbf{k}$ and not the moment vector $R\mathbf{k}$ itself, which is a vector pointing outward and normal to the XY plane (for $R > 0$).

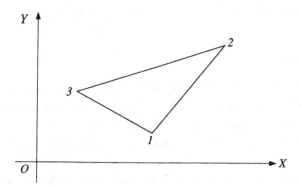

Figure 2.3 Triangle in the XY plane.

The line bound vector {L, M; R} was elegantly represented by Grassmann by the three 2×2 determinants of the matrix

$$\begin{bmatrix} 1 & x_1 & y_1 \\ 1 & x_2 & y_2 \end{bmatrix},$$

obtained by deleting, in turn, the columns

$$\begin{bmatrix} y_1 \\ y_2 \end{bmatrix}, \begin{bmatrix} x_1 \\ x_2 \end{bmatrix}, \text{ and } \begin{bmatrix} 1 \\ 1 \end{bmatrix}.$$

Thus

$$L = \begin{vmatrix} 1 & x_1 \\ 1 & x_2 \end{vmatrix}, \quad M = \begin{vmatrix} 1 & y_1 \\ 1 & y_2 \end{vmatrix}, \quad R = \begin{vmatrix} x_1 & y_1 \\ x_2 & y_2 \end{vmatrix}.$$

Furthermore, the area of a triangle with vertices 1, 2, and 3 and coordinates (x_1, y_1), (x_2, y_2), and (x_3, y_3) (see Fig. 2.3) is given by

$$\Delta = \frac{1}{2} \begin{vmatrix} 1 & x_1 & y_1 \\ 1 & x_2 & y_2 \\ 1 & x_3 & y_3 \end{vmatrix}. \tag{2.9}$$

Therefore, the area of the triangle $O12$ (see Fig. 2.1) is given by

$$\Delta = \frac{1}{2} \begin{vmatrix} 1 & x_1 & y_1 \\ 1 & x_2 & y_2 \\ 1 & 0 & 0 \end{vmatrix} = \frac{1}{2} \begin{vmatrix} x_1 & y_1 \\ x_2 & y_2 \end{vmatrix} = \frac{1}{2} R. \tag{2.10}$$

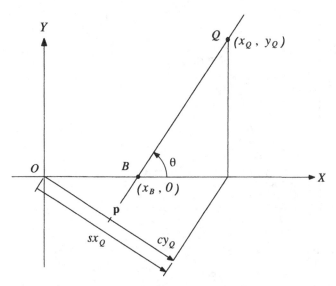

Figure 2.4 Determination of **p**.

Therefore, R is twice the area of the triangle $O12$ taken in the sense $O, 1, 2$. From Figure 2.1:

$$\Delta = \frac{1}{2} p \, |S| = \frac{1}{2} p \, (L^2 + M^2)^{1/2}, \tag{2.11}$$

where p is the length of the vector $\mathbf{p}$ drawn from O perpendicular to the line. Comparing (2.10) and (2.11) yields

$$p = \frac{R}{(L^2 + M^2)^{1/2}}. \tag{2.12}$$

Therefore, for a unit line segment for which $|S| = (L^2 + M^2)^{1/2} = 1$ we have $R = p$ and the line coordinates are $\{c, s; p\}$, with the abbreviations $c = \cos \theta$ and $s = \sin \theta$. If we substitute these results into (2.6), the equation for the line can be expressed in the form

$$cy - sx + p = 0. \tag{2.13}$$

If the coordinates of any point (x_Q, y_Q) or $(x_B, 0)$ on the line are known together with θ (see Fig. 2.4), the value for p is easy to obtain from (2.13)

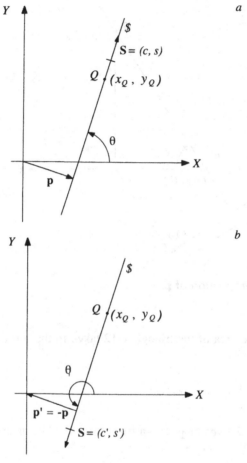

Figure 2.5 Determination of the line coordinates.

and $p = sx_Q - cy_Q = sx_B$. These expressions for p are easy to verify by simple projection using Figure 2.4.

The coordinates for a line can be measured in two distinct ways. The same line $ is illustrated in Figure 2.5a and b.

The homogeneous coordinates for the line using Figure 2.5a are $\{\cos\theta, \sin\theta; p\}$, where $p = x_Q \sin\theta - y_Q \cos\theta$, whereas the homogeneous coordinates for the line using part b are $\{\cos\theta', \sin\theta'; p'\}$. Now, $\theta' = \theta + \pi$ and, hence, $\cos\theta' = -\cos\theta$, $\sin\theta' = -\sin\theta$, and $p' = (x_Q \sin\theta' - y_Q \cos$

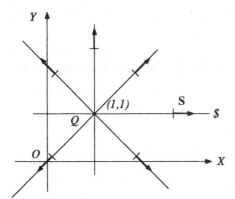

Figure 2.6 Generation of a pencil through the point with coordinates (1, 1).

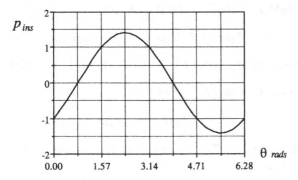

Figure 2.7 Plot of p versus θ.

$\theta') = -(x_Q \sin \theta - y_Q \cos \theta) = -p$. Thus, the homogeneous coordinates for the line $\$$ using part b of Figure 2.5 are $\{-\cos \theta, -\sin \theta; -p\}$, which are the same as $\{\cos \theta, \sin \theta, p\}$. It is, however, important to recognize that p can assume positive or negative values.

Assigning a sign to p is easily done by inspection. In part a of Figure 2.5 the unit vector $\mathbf{S}$ is turning about O in an anticlockwise sense and so p is a positive quantity, whereas in part b the unit vector $\mathbf{S}'$ is turning about O in a clockwise sense and, hence, p is a negative quantity.

It is interesting to plot p versus θ, and this is shown in Figure 2.6 for the pencil that passes through the point Q with coordinates (1,1) in Figure 2.7.

Consider that the pencil is formed by rotating the line labeled $ about Q with the vector **S** attached to it.

2.2 The coordinates for the point of intersection of a pair of lines

Consider the pair of lines

$$L_1 y - M_1 x + R_1 = 0 \tag{2.14}$$

and

$$L_2 y - M_2 x + R_2 = 0. \tag{2.15}$$

Then eliminate y by subtracting L_1 times (2.15) from L_2 times (2.14), which yields

$$(L_1 M_2 - L_2 M_1)x + (R_1 L_2 - R_2 L_1) = 0, \quad \text{or}$$

$$(L_1 M_2 - L_2 M_1)x = (L_1 R_2 - L_2 R_1). \tag{2.16}$$

Eliminate x by subtracting M_1 times (2.15) from M_2 times (2.14), which yields

$$(L_1 M_2 - L_2 M_1)y + (R_1 M_2 - R_2 M_1) = 0, \quad \text{or}$$

$$(L_1 M_2 - L_2 M_1)y = (M_1 R_2 - M_2 R_1). \tag{2.17}$$

It follows from (2.16) and (2.17) that the coordinates for the point of intersection can be expressed by

$$y : x : 1 = \begin{vmatrix} M_1 & R_1 \\ M_2 & R_2 \end{vmatrix} : \begin{vmatrix} L_1 & R_1 \\ L_2 & R_2 \end{vmatrix} : \begin{vmatrix} L_1 & M_1 \\ L_2 & M_2 \end{vmatrix}. \tag{2.18}$$

The sequence of determinants on the right side of (2.18) can be obtained directly from the matrix of line coordinates using Grassmann's expansion of the 2×2 determinants of

$$\begin{bmatrix} L_1 & M_1 & R_1 \\ L_2 & M_2 & R_2 \end{bmatrix},$$

by deleting, in turn, the columns

$$\begin{bmatrix} L_1 \\ L_2 \end{bmatrix}, \begin{bmatrix} M_1 \\ M_2 \end{bmatrix} \text{ and } \begin{bmatrix} R_1 \\ R_2 \end{bmatrix}.$$

The ratios of the coordinates of the point of intersection of a pair of lines with normalized coordinates $\{c_1, s_1; p_1\}$ and $\{c_2, s_2; p_2\}$ can be obtained directly from the matrix

$$\begin{bmatrix} c_1 & s_1 & p_1 \\ c_2 & s_2 & p_2 \end{bmatrix},$$

and

$$y : x : 1 = \begin{vmatrix} s_1 & p_1 \\ s_2 & p_2 \end{vmatrix} : \begin{vmatrix} c_1 & p_1 \\ c_2 & p_2 \end{vmatrix} : \begin{vmatrix} c_1 & s_1 \\ c_2 & s_2 \end{vmatrix}. \tag{2.19}$$

EXERCISE 2.1

1. On separate figures draw the lines that join the pairs of points $(1,1)$, $(-2,-2)$; $(4,1)$, $(-2,-3)$; $(-1,3)$, $(3,2)$; and $(-2,2)$, $(1,-5)$.

Using Grassmann's expansion of the 2×3 matrix

$$\begin{bmatrix} 1 & x_1 & y_1 \\ 1 & x_2 & y_2 \end{bmatrix}$$

determine the two sets of Plücker coordinates $\{L, M; R\}$ for each line obtained by interchanging the rows of the matrix. Label each line with the pair of equations $Ly - Mx + R = O$ and compare your results. In each figure draw the triangle O, 1, 2 and determine the signed area $O12$ using (2.10). Compare these values with the corresponding values of R you have obtained. Express each set of Plücker coordinates in the unitized form $\{c, s; p\}$, and write the pair of equations for each line. Label each line with the directed angle θ and the directed perpendicular distance p.

Using Grassmann's expansion of the 2×3 matrix

$$\begin{bmatrix} L_1 & M_1 & R_1 \\ L_2 & M_2 & R_2 \end{bmatrix}$$

determine the coordinates of the intersections of the pairs of lines $\{3.03, 1.75; 3.5\}$, $\{-1.03, 2.82; -6\}$; and $\{-2.82, 1.03; -2.1\}$, $\{-1, 1.73; 3.4\}$. Verify your results by drawing each pair of lines.

2.3   The statics of plane rigid systems

The concepts developed in the previous section can now be applied directly to the statics of planar rigid systems. The directed line segment **S** can be considered equivalent to a force applied to a rigid lamina (see Fig.

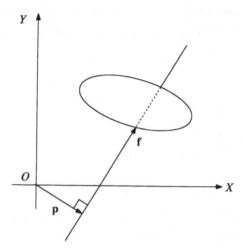

Figure 2.8 Representation of a force on a rigid lamina.

2.8). Because the lamina is rigid, the point of application can be moved any-where along the line. The first two coordinates of the directed line segment $\{L, M; R\}$ are called the *components of the force*, and the length of the line segment $(L^2 + M^2)^{1/2}$ is the *magnitude of the force*. From (2.12) the turning moment R is the product of the distance p and the magnitude of the force.

The problem of determining the resultant of an arbitrary system of forces $\{L_i, M_i; R_i\}$, $i = 1, 2, \ldots n$, that acts upon a plane lamina is essentially that of determining a unique line bound vector with the coordinates

$$L = \sum_{i=1}^{n} L_i, \quad M = \sum_{i=1}^{n} M_i, \quad R = \sum_{i=1}^{n} R_i. \tag{2.20}$$

However, there is an important exception. Consider the resultant of a pair of equal and opposite forces with the coordinates $\{L, M; R_1\}$ and $\{-L, -M; R_2\}$, where $R_1 \neq -R_2$ (see Fig. 2.9). The coordinates of the resultant $\{0, 0; R_1 + R_2\}$ are not a line bound vector, but a *pure couple*. Clearly, $(L^2 + M^2)^{1/2} = 0$ and from (2.12), $p = \infty$. As illustrated by Figure 2.9, a couple can be considered a force of infinitesimal magnitude, $|\delta f| = (L^2 + M^2)^{1/2} \to 0$, acting along a line which is parallel to the lines of action of the pair of forces and is infinitely distant, $p = \infty$, such that $|\delta f| p = (R_1 + R_2)$. This line is called

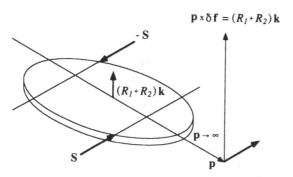

$$\mathbf{p} \times \delta \mathbf{f} = (R_1 + R_2) \mathbf{k}$$

Figure 2.9 A pure couple.

the *line at infinity*,* and the coordinates of the resultant couple $\{0, 0; R_1 + R_2\}$ can be expressed as $(R_1 + R_2) \{0, 0; 1\}$, where $(R_1 + R_2)$ is the magnitude of the resultant and $\{0, 0; 1\}$ are the coordinates of the line at infinity in the XY plane. A pure couple can thus be represented as a scalar multiple of the line of infinity, and hence there is no exception to the addition of forces in the XY plane. This representation is completely compatible with the fact that $p \times \delta \mathbf{f} = (R_1 + R_2) \mathbf{k}$, which is a free vector associated with the direction $\mathbf{k}$ and not with a definite line.

It is of interest to pursue this example by considering a pair of antiparallel forces that act upon a rigid lamina (i.e., a pair of forces which acts on parallel lines but in the opposite sense). We assume, without loss of generality, that the lines of action of the forces are parallel to the Y axis (see Fig. 2.10) and their coordinates are thus given by $\{0, M_1; R_1\}$ and $\{0, -M_2; -R_2\}$, or $M_1\{0, 1; p_1\}$ and $-M_2\{0, 1, p_2\}$, where $M_1 \neq 0$, and $M_2 \neq 0$.

The magnitude of the resultant is given by

$$M = M_1 - M_2. \tag{2.21}$$

---

* For more details on the line at infinity, see Klein (1939) and Hunt (1990). Briefly, in the Euclidean plane a pair of parallel lines do not meet. This is an exception to the general statement, "the meet of two lines is a point and the join of two points is a line." Introducing the line at infinity overcomes this difficulty and pairs of parallel lines meet at points at infinity, all of which lie on the line at infinity.

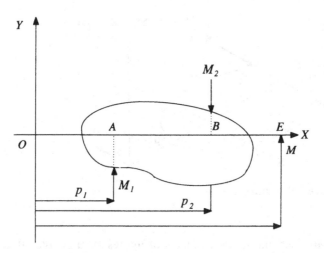

Figure 2.10 The resultant of a pair of antiparallel forces.

Furthermore,

$$Mp = M_1 p_1 - M_2 p_2,$$ (2.22)

and therefore the line of action of the resultant can be denoted by its distance p from O,

$$p = \frac{M_1}{M_1 - M_2} p_1 - \frac{M_2}{M_1 - M_2} p_2.$$ (2.23)

The substitution of $\lambda = M_1/M_2$ in (2.23) yields

$$p = \frac{\lambda}{\lambda - 1} p_1 - \frac{1}{\lambda - 1} p_2.$$ (2.24)

In Figure 2.10 the rigid lamina is represented by a contour drawn on the XY plane, and the lines of action of the applied antiparallel forces pass through points A and B. Let E be the point where the line of action of the resultant meets the X axis. The combined effect of applying the antiparallel forces upon the lamina is thus equivalent to extending the boundary of the lamina and applying a single resultant force through the point E. In this sense, the lamina is considered to be unbounded.

Assume that $M_2 > M_1$, and M_2 increases. The line of action of the resul-

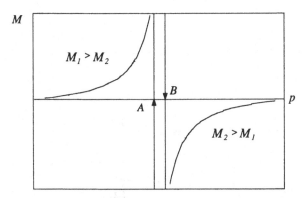

Figure 2.11 Plot of M versus p.

tant will approach B. From (2.24), when $M_2 \to \infty$, then $\lambda = 0$ and the line of action of M becomes $p = p_2$, and thus the resultant passes through B. Additionally, when $M_1 = M_2$, then $\lambda = 1$, and from (2.24) $p \to \infty$, and the line of action of the resultant lies on the line at infinity. The M versus p plot is the rectangular hyperbola labeled $M_2 > M_1$, as illustrated in Figure 2.11.

Assume now that $M_1 > M_2$. It is convenient to express (2.24) in the form

$$p = \frac{1}{1 - (1/\lambda)} p_1 - \frac{(1/\lambda)}{1 - (1/\lambda)} p_2. \tag{2.25}$$

As M_1 increases, the line of action of the resultant will approach A. From (2.25) when $M_1 \to \infty$, then $1/\lambda = 0$ and $p = p_1$. The line of action of the resultant thus passes through A. Furthermore, when $M_2 = M_1$, $\lambda = 1$, from (2.25) $p \to \infty$, and the line of action of the resultant lies on the line at infinity. The M versus p plot is the rectangular hyperbola labeled $M_1 > M_2$, as illustrated in Figure 2.11.

The resultant force thus lies to the left of point A or to the right side of point B, and it may not physically act upon the lamina. The forces at A and B are thus equivalent to a single force that acts upon a line as if the lamina were extended to include the line of action of this resultant force. Clearly, the resultant could lie on the line at infinity, and in statics it follows that the lamina has no boundary, but rather it can be considered infinitely large.

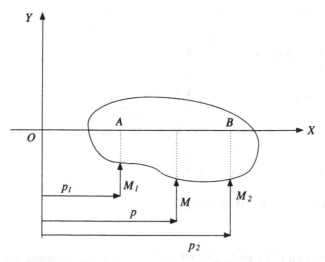

Figure 2.12 The resultant of a pair of parallel forces.

The magnitude of the resultant of a pair of parallel forces $\{0, M_1; R_1\}$, $\{0, M_2; R_2\}$ or $M_1\{0, 1; p_1\}$ and $M_2\{0, 1; p_2\}$ (see Fig. 2.12) is given by

$$M = M_1 + M_2. \tag{2.26}$$

The line of action of the resultant is given by

$$Mp = M_1p_1 + M_2p_2,$$

and therefore

$$p = \frac{M_1}{(M_1 + M_2)}p_1 + \frac{M_2}{(M_1 + M_2)}p_2. \tag{2.27}$$

It is a simple matter to divide the quotients on the right side of (2.27) above and below by M_1 and then by M_2, and to deduce that as $M_1 \to \infty$, $p \to p_1$ and as $M_2 \to \infty$, $p \to p_2$. Also, when $M_1 = M_2$, $p = \frac{1}{2}(p_1 + p_2)$. The M versus p plot is illustrated by Figure 2.13.

It is interesting to note that the resultants of antiparallel and parallel forces lie on pencils of parallel lines. In addition, the resultant of a pair of intersecting forces with coordinates $\{L_1, M_1; 0\}$, $\{L_2, M_2; 0\}$ (see Fig. 2.14)

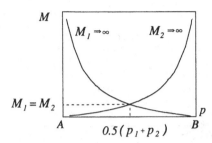

Figure 2.13 Plot of M versus P.

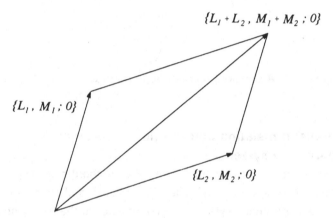

Figure 2.14 The resultant of a pair of intersecting forces.

lies on a pencil of lines that passes through the point of intersection, 0 (see Fig. 2.15).

In the three cases illustrated by Figures 2.10, 2.12, and 2.14, it is clear that the resultant is a linear combination of a pair of forces. In other words, any force in a pencil of forces is linearly dependent on any pair of forces in the pencil which does not lie on the same line. It is important to recognize that any force which does not belong to a particular pencil cannot be expressed as a linear combination of any number of forces in the pencil. Such a force is linearly independent of the set of forces which constitutes the pencil.

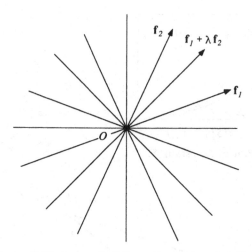

Figure 2.15 A pencil of lines through point *O*.

2.4   Parallel translation and rotation of rectangular coordinate systems

Assume that the two points 1 and 2 with coordinates (x_1, y_1) and (x_2, y_2) have the coordinates (x'_1, y'_1) and (x'_2, y'_2) before and after some transformation of the coordinate system. The coordinates of the line bound vector that join points 1 and 2 before and after the transformation are, respectively,

$$L = x_2 - x_1, \quad M = y_2 - y_1, \quad R = x_1 y_2 - x_2 y_1, \tag{2.28}$$

and

$$L' = x'_2 - x'_1, \quad M' = y'_2 - y'_1, \quad R' = x'_1 y'_2 - x'_2 y'_1. \tag{2.29}$$

2.4.1 Parallel translation of a rectangular coordinate system

The translation is illustrated by Figure 2.16, and it is clear that

$$x_1 = x'_1 + a, \quad x_2 = x'_2 + a,$$
$$y_1 = y'_1 + b, \quad y_2 = y'_2 + b. \tag{2.30}$$

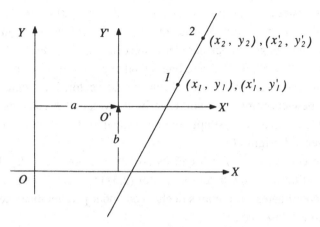

Figure 2.16 A parallel translation of a rectangular coordinate system.

It should also be apparent that the points 1 and 2 do not change under this transformation. Therefore, neither does the directed line segment 12 (or force) nor the line joining the points 1 and 2. All of these geometric elements, point, directed line segment or force, and unlimited line, are therefore invariant with the translation of the coordinate system. Substituting (2.30) in (2.29) and comparing with (2.28) yields

$$L = L', \quad M = M', \quad R = R' - L'b + M'a, \tag{2.31}$$

which can be expressed in matrix form as

$$\begin{bmatrix} L \\ M \\ R \end{bmatrix} = \begin{bmatrix} 1 & 0 & 0 \\ 0 & 1 & 0 \\ -b & a & 1 \end{bmatrix} \begin{bmatrix} L' \\ M' \\ R' \end{bmatrix}. \tag{2.32}$$

The matrix equation (2.32) expresses the coordinates of the force in the old coordinate system in terms of the coordinates of the force in the new system.

Thus far, coordinates have been represented by $\{L, M; R\}$. These coordinates now appear as the column vector on the left side of (2.32) and are represented by

$$\begin{bmatrix} L \\ M \\ R \end{bmatrix}.$$

The two coordinates (L, M) remain unchanged under translation and are therefore independent of the coordinate system. They define a free vector and their values remain unchanged under parallel displacement. As stated earlier, the three coordinates $\{L, M; R\}$ define a line bound vector. The coordinates $\{L', M'; R'\}$ define precisely the *identical* line bound vector, line segment (or force) in the new coordinate system. Clearly, the value of R changes and is origin dependent. Also, R' is simply the quantification of the moment of (L, M) about a second origin O'.

For a pure couple, $L = M = 0$ and its coordinates are $\{0, 0; R\}$. Let us make this substitution in the latter relation of (2.31) to yield $R = R'$. A *pure couple* with coordinates $R\mathbf{k}$ remains unchanged under parallel displacements and is therefore a *free vector*.

We can now give a more direct derivation of the transformation matrix in (2.32). The transformation is of the form

$$
\begin{bmatrix} L \\ M \\ R \end{bmatrix} = \begin{bmatrix} a_{11} & a_{12} & a_{13} \\ a_{21} & a_{22} & a_{23} \\ a_{31} & a_{32} & a_{33} \end{bmatrix} \begin{bmatrix} L' \\ M' \\ R' \end{bmatrix}. \tag{2.33}
$$

The line coordinates of the X' and Y' axes in the $O'\,X'\,Y'$ coordinate system are represented in matrix form by the column vectors

$$
\begin{bmatrix} 1 \\ 0 \\ 0 \end{bmatrix} \quad \text{and} \quad \begin{bmatrix} 0 \\ 1 \\ 0 \end{bmatrix}.
$$

The moments of the X' and Y' axes about O are, respectively (see Fig. 2.17), by $\mathbf{r} \times \mathbf{i} = -b\mathbf{k}$ and $\mathbf{r} \times \mathbf{i} = a\mathbf{k}$, where $\mathbf{r} = a\mathbf{i} + b\mathbf{i}$. Therefore, the column vectors that represent the X' and Y' axes in the OXY system are respectively

$$
\begin{bmatrix} 1 \\ 0 \\ -b \end{bmatrix} \quad \text{and} \quad \begin{bmatrix} 0 \\ 1 \\ a \end{bmatrix}.
$$

These coordinates can be obtained directly from an inspection of Figure 2.17: The direction cosines for the X' and Y' axes are $(1,0)$ and $(0,1)$, respectively, while their moments about O are $-b$ and a, respectively.

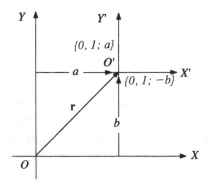

Figure 2.17 Line coordinates for the X' and Y' axes.

The substitution of these results into the left and right sides of (2.33) yields

$$\begin{bmatrix} 1 \\ 0 \\ -b \end{bmatrix} = \begin{bmatrix} a_{11} & a_{12} & a_{13} \\ a_{21} & a_{22} & a_{23} \\ a_{31} & a_{32} & a_{33} \end{bmatrix} \begin{bmatrix} 1 \\ 0 \\ 0 \end{bmatrix}, \tag{2.34}$$

and

$$\begin{bmatrix} 0 \\ 1 \\ a \end{bmatrix} = \begin{bmatrix} a_{11} & a_{12} & a_{13} \\ a_{21} & a_{22} & a_{23} \\ a_{31} & a_{32} & a_{33} \end{bmatrix} \begin{bmatrix} 0 \\ 1 \\ 0 \end{bmatrix}. \tag{2.35}$$

It follows from (2.34) and (2.35) that the first and second columns of the transformation matrix are, respectively,*

$$\begin{bmatrix} a_{11} \\ a_{21} \\ a_{31} \end{bmatrix} = \begin{bmatrix} 1 \\ 0 \\ -b \end{bmatrix}, \tag{2.36}$$

---

* There is a well-known result in linear algebra which states that the column vectors of a matrix of a linear transformation represents the image of the basis vectors

$$\begin{bmatrix} 0 \\ 0 \\ 1 \end{bmatrix}, \begin{bmatrix} 0 \\ 1 \\ 0 \end{bmatrix}, \quad \text{and} \quad \begin{bmatrix} 0 \\ 0 \\ 1 \end{bmatrix}$$

in terms of the new basis.

and

$$\begin{bmatrix} a_{12} \\ a_{22} \\ a_{32} \end{bmatrix} = \begin{bmatrix} 0 \\ 1 \\ a \end{bmatrix}. \tag{2.37}$$

The first two columns of the transformation matrix are the line coordinates for the X' and Y' axes in the old coordinate system. The third column is the line coordinate for the line at infinity, which remains unchanged under any transformation:

$$\begin{bmatrix} 0 \\ 0 \\ 1 \end{bmatrix} = \begin{bmatrix} a_{11} & a_{12} & a_{13} \\ a_{21} & a_{22} & a_{23} \\ a_{31} & a_{32} & a_{33} \end{bmatrix} \begin{bmatrix} 0 \\ 0 \\ 1 \end{bmatrix}, \tag{2.38}$$

and, hence,

$$\begin{bmatrix} a_{13} \\ a_{23} \\ a_{33} \end{bmatrix} = \begin{bmatrix} 0 \\ 0 \\ 1 \end{bmatrix}. \tag{2.39}$$

These results yield (2.32) precisely.

2.4.2 Rotation of a rectangular coordinate system

Assume that the $X'Y'$ coordinate system is rotated through an angle ϕ, as illustrated by Figure 2.18. From the figure:

$$x' = r' \cos(\alpha + \phi) = r' \cos \alpha \cos \phi - r' \sin \alpha \sin \phi,$$
$$y' = r' \sin(\alpha + \phi) = r' \sin \alpha \cos \phi + r' \cos \alpha \sin \phi, \tag{2.40}$$

and

$$x'' = r' \cos \alpha, \quad y'' = r' \sin \alpha. \tag{2.41}$$

The substitution of (2.41) in (2.40) yields the two expressions for the rotation

$$x' = x'' \cos \phi - y'' \sin \phi,$$
$$y' = x'' \sin \phi + y'' \cos \phi. \tag{2.42}$$

Points 1 and 2 (see Fig. 2.19) do not change under this transformation and

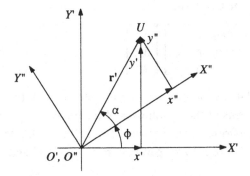

Figure 2.18 A rotation of a rectangular coordinate system.

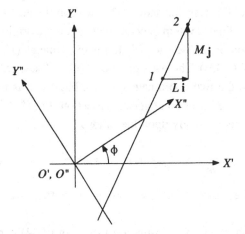

Figure 2.19 Transformation of line coordinates due to a rotation of the coordinate system.

neither does the directed line segment 12 (or force) nor the line joining the points 1 and 2. All these geometric elements are therefore invariant with a rotation of coordinate system. The substitution of (2.42) into $L' = x_2' - x_1'$, $M' = y_2' - y_1'$, and $R' = x_1'y_2' - x_2'y_1'$ yields

$$L' = L'' \cos \phi - M'' \sin \phi,$$

$$M' = L'' \sin \phi + M'' \cos \phi, \tag{2.43}$$

$$R' = R'',$$

which can be expressed in matrix form as

$$
\begin{bmatrix} L' \\ M' \\ R' \end{bmatrix} = \begin{bmatrix} \cos\phi & -\sin\phi & 0 \\ \sin\phi & \cos\phi & 0 \\ 0 & 0 & 1 \end{bmatrix} \begin{bmatrix} L'' \\ M'' \\ R'' \end{bmatrix}. \tag{2.44}
$$

The matrix equation (2.44) expresses the coordinates of the force in the old coordinate system in terms of the coordinates of the force in the new co-ordinate system. The first two relations between (L', M') and (L'', M'') can be obtained directly from Figure 2.19 by parallel projection of L' and M' on the X'' and Y'' axes.

For this transformation of coordinates, R remains unchanged. This is not a surprising result because the axis of rotation, the Z axis pointing outward through O, is invariant under this transformation. The two coordinates (L'', M'') of the vector in the new coordinate system depend solely upon the original values (L', M'), and the relation of (L', M') to (L'', M'') is the same as (x', y') to (x'', y'') (compare (2.42) and (2.43)). The pairs of values, (L'', M'') and (L', M'), quantify the same vector in the two coordinate systems. Finally, it is evident from (2.43) that $\{L', M'; R'\}$ are linear homogeneous functions of $\{L'', M''; R''\}$, and the ratios $L' : M' : R'$ depend solely upon the ratios $L'' : M'' : R''$:

$$
\frac{L'}{L''} = \cos\phi, \ -\frac{M''}{L''}\sin\phi, \ \frac{M'}{M''} = \frac{L''}{M''}\sin\phi + \cos\phi, \ \frac{R'}{R''} = 1. \tag{2.45}
$$

We have established that these three ratios (without regard to their actual values) determine the line.

A more direct derivation of the transformation matrix in (2.44) can now be given. The transformation takes the form

$$
\begin{bmatrix} L' \\ M' \\ R' \end{bmatrix} = \begin{bmatrix} a_{11} & a_{12} & a_{13} \\ a_{21} & a_{22} & a_{23} \\ a_{31} & a_{32} & a_{33} \end{bmatrix} \begin{bmatrix} L'' \\ M'' \\ R'' \end{bmatrix}. \tag{2.46}
$$

By analogy with (2.36) and (2.37), the first two columns of the matrix are the line coordinates of the X'', Y'' axes in the X', Y' system, which by inspection of Figure 2.19 are

$$
\begin{bmatrix} a_{11} \\ a_{21} \\ a_{31} \end{bmatrix} = \begin{bmatrix} \cos\phi \\ \sin\phi \\ 0 \end{bmatrix}, \tag{2.47}
$$

and

$$\begin{bmatrix} a_{12} \\ a_{22} \\ a_{32} \end{bmatrix} = \begin{bmatrix} \cos(\phi + \pi/2) \\ \sin(\phi + \pi/2) \\ 0 \end{bmatrix} = \begin{bmatrix} -\sin \phi \\ \cos \phi \\ 0 \end{bmatrix}. \tag{2.48}$$

By analogy with (2.39), the third column consists of the line coordinates for the line at infinity,

$$\begin{bmatrix} a_{13} \\ a_{23} \\ a_{33} \end{bmatrix} = \begin{bmatrix} 0 \\ 0 \\ 1 \end{bmatrix}. \tag{2.49}$$

2.4.3 The Euclidean group of motions

All combinations of linear translations in the plane and rotations around the Z axis constitute the group of Euclidean motions in the plane, which, upon combining (2.30) without the subscripts 1 or 2 and (2.42), can be expressed in the form

$$x = x'' \cos \phi - y'' \sin \phi + a,$$
$$y = x'' \sin \phi + y'' \cos \phi + b, \tag{2.50}$$

or in matrix form as

$$\begin{bmatrix} x \\ y \\ 1 \end{bmatrix} = \begin{bmatrix} \cos \phi & -\sin \phi & a \\ \sin \phi & \cos \phi & b \\ 0 & 0 & 1 \end{bmatrix} \begin{bmatrix} x'' \\ y'' \\ 1 \end{bmatrix}. \tag{2.51}$$

It is probably self-evident to the reader that scalar geometrical configurations, such as the distance between a pair of points, the angle between a pair of lines, and the area of a triangle, must remain invariant under the Euclidean group of motions. We have shown in previous sections that directed line segments (line bound vectors) which are geometrically equivalent to a force, free vectors such as couples which are geometrically equivalent to area, and the line, all remain invariant with the Euclidean group of motions. We can consider such quantities meaningful geometrical configurations in precisely the same way as the scalar quantities: distance, angle, and area.

These considerations stem from the two fundamental principles of Grassmann (see Klein 1939).

First Grassmann Principle. The geometric properties of any figures must be expressible in formulas which are not changed under a transformation of the coordinate system. Conversely, any formula which, in this sense, is invariant under coordinate transformations must represent a geometric property.

A coordinate system is clearly arbitrary (noninvariant in the previous sense) and does not represent a geometric property.

Second Grassmann Principle. If the system of magnitudes such as L', M', and R' formed from the transformed coordinates of the points 1 and 2 expresses itself exclusively in terms of the magnitudes L'', M'', and R'' formed in the same way from the initial coordinates (the coordinates themselves do not appear explicitly), then the system defines a geometric configuration, i.e., one which is independent of the coordinate system. In fact, all analytic expressions can be classified according to their behavior under coordinate transformations, and two expressions which transform in the same way (such as expressions for area and turning moment) are defined as geometrically equivalent.

2.5 Induced force/line transformation under the Euclidean group

From (2.32) and (2.44), the coordinates for a force in the old coordinate system expressed in terms of the coordinates of the force in the new coordinate system for both a translation and rotation are given by

$$\begin{bmatrix} L \\ M \\ R \end{bmatrix} = [e] \begin{bmatrix} L'' \\ M'' \\ R'' \end{bmatrix}, \tag{2.52}$$

where

$$[e] = \begin{bmatrix} 1 & 0 & 0 \\ 0 & 1 & 0 \\ -b & a & 1 \end{bmatrix} \begin{bmatrix} c & -s & 0 \\ s & c & 0 \\ 0 & 0 & 1 \end{bmatrix}$$

$$= \begin{bmatrix} c & -s & 0 \\ s & c & 0 \\ as - bc & ac + bs & 1 \end{bmatrix}, \tag{2.53}$$

and the abbreviations are $c = \cos \phi$ and $s = \sin \phi$, as introduced earlier.

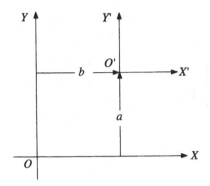

Exercise Figure 2.2(1)

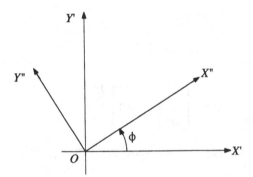

Exercise Figure 2.2(2)

──────── **EXERCISE 2.2** ────────

1. Write the Plücker coordinates for the X' and Y' axes in the OXY coordinate system and the Plücker coordinates for the X and Y axes in the $O'X'Y'$ coordinate system (see Exr. Fig. 2.2(1)). Determine the matrices $[A]$ and $[A']$ for the induced force transformations

$$\begin{bmatrix} L \\ M \\ R \end{bmatrix} = [A] \begin{bmatrix} L' \\ M' \\ R' \end{bmatrix} \quad \text{and} \quad \begin{bmatrix} L' \\ M' \\ R' \end{bmatrix} = [A'] \begin{bmatrix} L \\ M \\ R \end{bmatrix}$$

for pure translations of the coordinate systems. Show that $[A'] = [A]^{-1}$.

2. Write the Plücker coordinates for the X'' and Y'' axes in the $OX'Y'$ coordinate system and the Plücker coordinates for the X', Y' axes in the $OX''Y''$ coordinate system (see Exr. Fig. 2.2(2)). Determine the matrices $[B']$ and $[B'']$ for the

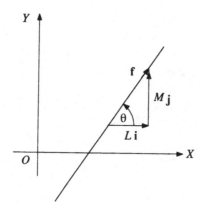

Figure 2.20 A vector representation of a force.

induced force transformations

$$\begin{bmatrix} L' \\ M' \\ R' \end{bmatrix} = [B'] \begin{bmatrix} L'' \\ M'' \\ R'' \end{bmatrix} \quad \text{and} \quad \begin{bmatrix} L'' \\ M'' \\ R'' \end{bmatrix} = [B''] \begin{bmatrix} L' \\ M' \\ R' \end{bmatrix}$$

for pure rotations of the coordinate systems. Show that $[B''] = [B']^{-1}$.
3. Determine $[e] = [A][B']$.

2.6   A useful vector representation of a force

We have established that the three coordinates $\{L, M; R\}$ represent a force with respect to some reference coordinate system, and the value for R is dependent on the choice of origin O. It is convenient to represent a force in the vector form with the coordinates (see Fig. 2.20)

$$\hat{w} = \{\mathbf{f}, \mathbf{c_o}\}. \tag{2.54}$$

The two representations are related by

$$\mathbf{f} = (L\mathbf{i} + M\mathbf{j}), \quad \mathbf{c_o} = R\mathbf{k}. \tag{2.55}$$

The subscript o is introduced to indicate that the moment vector is origin dependent. Furthermore, f can be expressed as a scalar multiple $f\mathbf{S}$ of a unit vector $\mathbf{S}$, $|\mathbf{S}| = 1$, where $f = |\mathbf{f}|$. Then the moment vector $\mathbf{c_o} = f\mathbf{r} \times \mathbf{S}$ can be expressed in the form $c_o = fS_o$, where $\mathbf{r} \times \mathbf{S} = \mathbf{S_o}$. The pair of vectors $(\mathbf{S}; \mathbf{S_o})$ thus determines a directed line segment of unit length, and they can be expressed by

$$\hat{s} = \{\mathbf{S}; \mathbf{S_o}\}. \tag{2.56}$$

The coordinates for a force can now be expressed by a scalar multiple f, the magnitude of the force, and a unit line vector:

$$\hat{w} = \{\mathbf{f}; \mathbf{c_o}\} = f\{\mathbf{S}; \mathbf{S_o}\} = f\hat{s}. \tag{2.57}$$

This latter representation separates the magnitude of the force f, from the geometric quantity $\hat{s}$.

2.7   The statics of a parallel manipulator

The upper lamina is connected to the fixed base via three parallel *RPR* kinematic chains (see Fig. 2.21). The prismatic pair in each chain is actuated and the platform has three degrees of freedom. Each prismatic pair clearly connects a pair of revolute joints, one embedded in the fixed frame and the other in the movable lamina. From here on, this prismatic pair will be called a connector, and each force generated in a connector will be called a connector force. We assume at the outset that the geometric configuration is known.

Consider that the three connector forces with magnitudes f_1, f_2, and f_3 are generated in each of the lines $\$_1$, $\$_2$, and $\$_3$ with the coordinates $\hat{s}_1$, $\hat{s}_2$, and $\hat{s}_3$. The upper lamina will experience a force of magnitude f acting on a line $\$$. Therefore,

$$\hat{w} = \hat{w}_1 + \hat{w}_2 + \hat{w}_3. \tag{2.58}$$

Equation 2.58 can be expressed in the alternative forms

$$\hat{w} = \begin{bmatrix} f \\ c_o \end{bmatrix} = \begin{bmatrix} f_1 \\ c_1 \end{bmatrix} + \begin{bmatrix} f_2 \\ c_2 \end{bmatrix} + \begin{bmatrix} f_3 \\ c_3 \end{bmatrix}. \tag{2.59}$$

and

$$\hat{w} = \begin{bmatrix} f \\ c_o \end{bmatrix} = f_1\hat{s}_1 + f_2\hat{s}_2 + f_3\hat{s}_3$$

$$= f_1 \begin{bmatrix} S_1 \\ S_{01} \end{bmatrix} + f_2 \begin{bmatrix} S_2 \\ S_{02} \end{bmatrix} + f_3 \begin{bmatrix} S_3 \\ S_{03} \end{bmatrix}. \tag{2.60}$$

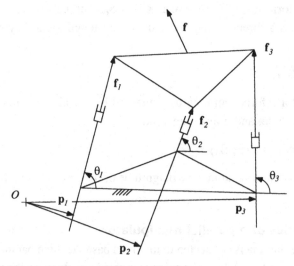

Figure 2.21 A parallel manipulator.

Therefore, from (2.60):

$$\mathbf{f} = f_1\mathbf{S}_1 + f_2\mathbf{S}_2 + f_3\mathbf{S}_3,$$ (2.61)

and

$$\mathbf{c}_o = f_1\mathbf{S}_{01} + f_2\mathbf{S}_{02} + f_3\mathbf{S}_{03}.$$ (2.62)

From (2.61):

$$f = |\mathbf{f}| = |f_1\mathbf{S}_1 + f_2\mathbf{S}_2 + f_3\mathbf{S}_3|,$$ (2.63)

and therefore

$$\mathbf{S} = \frac{f_1\mathbf{S}_1 + f_2\mathbf{S}_2 + f_3\mathbf{S}_3}{|f|}.$$ (2.64)

From (2.60) and (2.61):

$$\mathbf{S}_o = \frac{f_1\mathbf{S}_{01} + f_2\mathbf{S}_{02} + f_3\mathbf{S}_{03}}{|f|}$$ (2.65)

It is often convenient to express (2.60) in the expanded form

$$\hat{w} = f_1\begin{bmatrix} c_1 \\ s_1 \\ p_1 \end{bmatrix} + f_2\begin{bmatrix} c_2 \\ s_2 \\ p_2 \end{bmatrix} + f_3\begin{bmatrix} c_3 \\ s_3 \\ p_3 \end{bmatrix},$$ (2.66)

which can be expressed in the popular matrix form by

$$\hat{w} = j\lambda, \tag{2.67}$$

where

$$j = \begin{bmatrix} c_1 & c_2 & c_3 \\ s_1 & s_2 & s_3 \\ p_1 & p_2 & p_3 \end{bmatrix} \tag{2.68}$$

is a 3×3 matrix, and

$$\lambda = \begin{bmatrix} f_1 \\ f_2 \\ f_3 \end{bmatrix} \tag{2.69}$$

is a 3×1 column vector.

When connector forces of magnitudes f_1, f_2, and f_3 are generated for some specified configuration, the coordinates $\hat{w}$ of the resultant acting upon the moving platform can be computed from (2.67). Hence, the magnitude of the *resultant f* together with its line of action can be determined (see Fig. 2.22). This is called the *forward static analysis*. Clearly, *for equilibrium an external force of equal and opposite magnitude f* must be applied to the platform on the same line.

Conversely, when an external force with magnitude f on a line $ is applied to the moving platform we need to determine the *magnitudes f_1, f_2, and f_3 of the resultant connector forces* (see Fig. 2.23), which is easy to obtain from the equation

$$\lambda = j^{-1}\hat{w}, \tag{2.70}$$

where j^{-1} denotes the inverse of j. This is called the *reverse or inverse static analysis. For equilibrium, the equilibrant connector forces are equal and opposite.*

Assume now that a connector is an *RPR* kinematic chain joining a fixed and a movable platform at B_i and C_i (see Fig. 2.24) and that an internal force can be generated by a hydraulic system in the sliding joint. Furthermore, a force f_i is generated in the cylinder which acts upon the piston at point E_i. For equilibrium, an equal and opposite reaction force is transmitted from the movable platform to the connector at point C_i. The connector link C_iE_i is either in compression (see Fig. 2.24a) or in tension (see Fig. 2.24b). Analogously, the generated

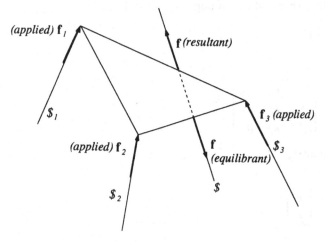

Figure 2.22 Forward analysis.

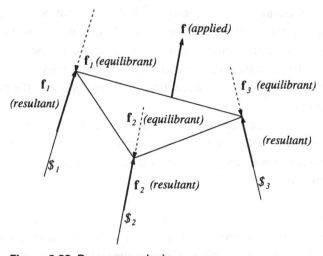

Figure 2.23 Reverse analysis.

force f_i acts upon the cylinder head at point H_i, and an equal and opposite re-action force is transmitted from the fixed platform at point B_i to the connector B_iH_i. The connector link B_iH_i is either in compression or tension, as illustrated in Figure 2.24. This is essentially a forward static analysis of a single connec-tor which identifies the states of compression and tension in a connector.

Conversely, when an external force f_i is applied at point C_i, then an equal

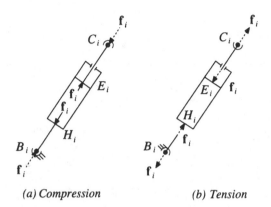

(a) Compression   (b) Tension

Figure 2.24 A static analysis of a single connector.

and opposite force f_i must be generated in the cylinder which acts at point E_i. The connector link C_iE_i can be either in compression or tension, as illustrated. Finally, when the external force f_i is applied at point C_i, an equal and opposite reaction force f_i is transmitted from the fixed platform at point B_i to the connector link B_iH_i, which can be in a state of compression or tension, as shown in Figure 2.24. This is essentially a reverse static analysis of a single connector which identifies the states of compression and tension in the connector.

―――― **EXERCISE 2.3** ――――――――――――――

1. (a) Determine the unitized coordinates of the lines that join the connectors B_1C_1, B_2C_2, and B_3C_3 using the point B_1 (B_2) as your reference (see Exr. Fig. 2.3(1)). Hence, determine (2.67), $\hat{w} = j\lambda$ for the truss.

 (b) A vertical force of 10 lbf is applied through point C_1. Compute the equilibrating connector forces f_1, f_2, and f_3, and state whether each connector is in compression or tension.

 (c) A vertical force of 10 lbf is applied through point C_2, and the force through point C_1 is removed. Compute the equilibrating connector forces and state whether each connector is in compression or tension. Confirm this result by choosing your reference point at C_2 (C_3).

2. (a) Determine the unitized coordinates for the lines that join the connectors B_1C_1, B_2C_2, and B_3C_3 using the point B_1 (B_2) as your reference (see Exr. Fig. 2.3(2)). Hence, determine (2.67), $\hat{w} = j\lambda$ for the truss.

 (b) A vertical force of 5 lbf is applied one inch to the right of point C_1.

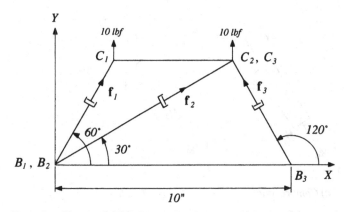

Exercise Figure 2.3(1)

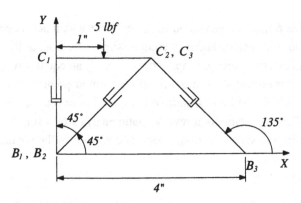

Exercise Figure 2.3(2)

Compute the equilibrating connector forces and state whether each connector is in compression or tension.

(c) Compute the equilibrating connector forces when the vertical force of 5 lbf is applied at C_1 and then at $C_2(C_3)$. Choose your reference points at $B_1(B_2)$, then $C_2(C_3)$.

2.8 The geometrical meaning of j^{-1}

The inverse of

$$j = \begin{bmatrix} c_1 & c_2 & c_3 \\ s_1 & s_2 & s_3 \\ p_1 & p_2 & p_3 \end{bmatrix}$$

can be determined by constructing the matrix,

$$j' = \begin{bmatrix} (s_2p_3 - s_3p_2) & (s_3p_1 - s_1p_3) & (s_1p_2 - s_2p_1) \\ -(c_2p_3 - c_3p_2) & -(c_3p_1 - c_1p_3) & -(c_1p_2 - c_2p_1) \\ (c_2s_3 - c_3s_2) & (c_3s_1 - c_1s_3) & (c_1s_2 - c_2s_1) \end{bmatrix}, \quad (2.71)$$

where the elements of j' are the signed minors of j. The inverse j^{-1} can thus be expressed in the form

$$j^{-1} = j'^{T}/\det j, \quad (2.72)$$

where

$$\det j = \begin{vmatrix} c_1 & c_2 & c_3 \\ s_1 & s_2 & s_3 \\ p_1 & p_2 & p_3 \end{vmatrix},$$

and j'^T denotes the transpose of j'.

The geometrical meaning of j^{-1} can be understood by solving sequentially the equations for the connector lines, which can expressed in the form

$$c_1y - s_1x + p_1 = 0, \quad (2.73)$$

$$c_2y - s_2x + p_2 = 0, \quad (2.74)$$

$$c_3y - s_3x + p_3 = 0. \quad (2.75)$$

The solution of (2.73) and (2.74) for the coordinates (x_{12}, y_{12}) of the point of intersection of the lines $\$_1$ and $\$_2$, using Grassmann's expansion (2.19), yields (see Fig. 2.25)

$$y_{12} : -x_{12} : 1 = (s_1p_2 - s_2p_1) : -(c_1p_2 - c_2p_1) : (c_1s_2 - c_2s_1). \quad (2.76)$$

Analogously,

$$y_{31} : -x_{31} : 1 = (s_3p_1 - s_1p_3) : -(c_3p_1 - c_1p_3) : (c_3s_1 - c_1s_3), \quad (2.77)$$

and

$$y_{23} : -x_{23} : 1 = (s_2p_3 - s_3p_2); -(c_2p_3 - c_3p_2) : (c_2s_3 - c_3s_2). \quad (2.78)$$

Chapter 3 will show that the array $\{y, -x, 1\}$ consists of the coordinates for a line $\$$ through a point with coordinates (x, y) and which is perpendicular to the XY plane. Hence, the columns in (2.71) are, to a scalar multiple,

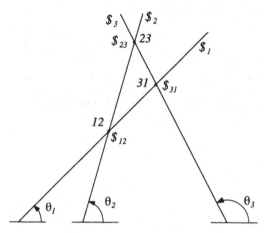

Figure 2.25 Skeleton form of a parallel manipulator illustrating the lines associated with the connectors.

the unitized homogeneous line coordinates of the lines $\$_{23}$, $\$_{31}$, and $\$_{12}$ through the points labeled 23, 13, and 12, which are perpendicular to the plane of the page (see Fig. 2.25). Therefore,

$$
\begin{bmatrix} (s_2 p_3 - s_3 p_2) \\ -(c_2 p_3 - c_3 p_2) \\ (c_2 s_3 - c_3 s_2) \end{bmatrix} = s_{3-2} \begin{bmatrix} y_{23} \\ -x_{23} \\ 1 \end{bmatrix} = s_{3-2}\, \hat{\$}_{23}, \tag{2.79}
$$

$$
\begin{bmatrix} (s_3 p_1 - s_1 p_3) \\ -(c_3 p_1 - c_1 p_3) \\ (c_3 s_1 - c_1 s_3) \end{bmatrix} = s_{1-3} \begin{bmatrix} y_{31} \\ -x_{31} \\ 1 \end{bmatrix} = s_{1-3}\, \hat{\$}_{31}, \tag{2.80}
$$

$$
\begin{bmatrix} (s_1 p_2 - s_2 p_1) \\ -(c_1 p_2 - c_2 p_1) \\ (c_1 s_2 - c_2 s_1) \end{bmatrix} = s_{2-1} \begin{bmatrix} y_{12} \\ -x_{12} \\ 1 \end{bmatrix} = s_{2-1}\, \hat{\$}_{12}, \tag{2.81}
$$

where the trigonometrical identity $c_j s_i - c_i s_j = s_{i-j}$ has been used. Furthermore, j^{-1} can be expressed in the abbreviated form

$$
j^{-1} = \begin{bmatrix} s_{3-2}\, \hat{\$}_{23}^{T} \\ s_{1-3}\, \hat{\$}_{31}^{T} \\ s_{2-1}\, \hat{\$}_{12}^{T} \end{bmatrix} / \det j. \tag{2.82}
$$

The rows of j^{-1} are therefore, to a scalar multiple, the unitized coordinates for the lines $\$_{23}$, $\$_{31}$, and $\$_{12}$.

The expansion of det j from the first column yields

$$\det j = c_1(s_2 p_3 - s_3 p_2) - s_1(c_2 p_3 - c_3 p_2) + p_1(c_2 s_3 - c_3 s_2), \quad (2.83)$$

which, from (2.79), can be expressed in the form

$$\det j = \hat{s}_1^T s_{3-2} \hat{S}_{23} = s_{3-2} \hat{S}_{23}^T \hat{s}_1. \quad (2.84)$$

Analogously, expansion from the second and third columns yields

$$\det j = \hat{s}_2^T s_{1-3} \hat{S}_{31}^T = s_{1-3} \hat{S}_{31}^T \hat{s}_2, \quad (2.85)$$

and

$$\det j = \hat{s}_3^T s_{2-1} \hat{S}_{12} = s_{2-1} \hat{S}_{12}^T \hat{s}_3. \quad (2.86)$$

The substitution of (2.82) into the right side of (2.70) yields

$$\lambda = \begin{bmatrix} f_1 \\ f_2 \\ f_3 \end{bmatrix} = \begin{bmatrix} s_{3-2} \hat{S}_{23} \\ s_{1-3} \hat{S}_{31} \\ s_{2-1} \hat{S}_{12} \end{bmatrix}^T \hat{w} \,/\, \det j. \quad (2.87)$$

Therefore, from (2.87) and (2.84)–(2.86):

$$f_1 = \hat{S}_{23}^T \hat{w} / \hat{S}_{23}^T \hat{s}_1, \quad (2.88)$$

$$f_2 = \hat{S}_{31}^T \hat{w} / \hat{S}_{31}^T \hat{s}_2, \quad (2.89)$$

$$f_3 = \hat{S}_{12}^T \hat{w} / \hat{S}_{12}^T \hat{s}_3. \quad (2.90)$$

The geometrical meaning of these expressions will be made clear in Section 4.4

2.9 Singularity configurations of a parallel manipulator

It is clear that when the rank of j is less than 3, then det $j = 0$ (see equation 2.68) and hence, j^{-1} does not exist. When this occurs, the manipulator is in a special or singularity configuration. It is well known that when

$$\det j = \begin{vmatrix} c_1 & c_2 & c_3 \\ s_1 & s_2 & s_3 \\ p_1 & p_2 & p_3 \end{vmatrix} = 0,$$

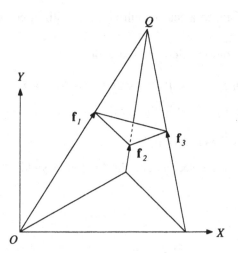

Figure 2.26 Singularity configuration; the connector forces are concurrent.

then the three lines are concurrent, i.e., they all meet in a finite or an infinite point. A configuration for Q finite is illustrated in Figure 2.26.

It is easy to deduce the det $j = 0$, for if the origin O of the XY coordinate system were transformed to Q, then $p_1 = p_2 = p_3 = 0$. Also, the lines of action of the forces pass through Q, and the three forces are thus linearly dependent (see Section 2.4). Consider that the origin for a coordinate system is located at Q and a force with coordinates $\hat{w} = (L, M; R)$ is applied to the top platform. It follows that

$$\begin{bmatrix} L \\ M \\ R \end{bmatrix} = \begin{bmatrix} c_1 & c_2 & c_3 \\ s_1 & s_2 & s_3 \\ 0 & 0 & 0 \end{bmatrix} \begin{bmatrix} f_1 \\ f_2 \\ f_3 \end{bmatrix}. \tag{2.91}$$

It is not possible to solve (2.91) for f_1, f_2, and f_3; clearly there is a couple, with magnitude R acting, which cannot be equilibrated by the connector forces generated. We will show in Section 4.6 that the top platform has an uncontrollable instant mobility, that is, an instantaneous rotation about Q.

A singular configuration occurs when all three connector forces are on parallel lines (or the point of intersection Q is at infinity). The geometry is

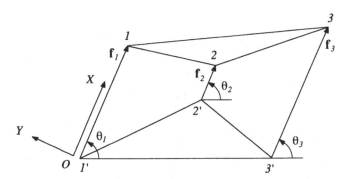

Figure 2.27 Singularity configuration; the connector forces are parallel.

illustrated in Figure 2.27. The slopes of the lines of action of the three forces are the same and equal to tan λ. Therefore,

$$\frac{s_1}{c_1} = \frac{s_2}{c_2} = \frac{s_3}{c_3} = \tan \lambda. \tag{2.92}$$

The first two rows of det j are identical to a scalar multiple of tan λ, and thus det $j = 0$. The three forces belong to a parallel pencil of lines and are linearly dependent (see Section 2.4). Consider that a coordinate system is chosen at a point O with the X axis drawn parallel to the legs, and a force $\hat{w} = \{L, M; R\}$ is applied to the top platform. It follows that

$$\begin{bmatrix} L \\ M \\ R \end{bmatrix} = \begin{bmatrix} 1 & 1 & 1 \\ 0 & 0 & 0 \\ 0 & p_2 & p_3 \end{bmatrix} \begin{bmatrix} f_1 \\ f_2 \\ f_3 \end{bmatrix}. \tag{2.93}$$

It is not possible to solve (2.93) for f_1, f_2, and f_3; clearly, any force parallel to the Y axis cannot be equilibrated by the connector forces generated. Section 4.6 shows that the top platform has an uncontrollable instant mobility, that is, an instantaneous translation parallel to the Y axis.

A singular position occurs when any two connector forces lie on the same line. Figure 2.28 illustrates the case for $\mathbf{f}_2 = \lambda \mathbf{f}_1$, where $\lambda \neq 0$. (Clearly, two more cases occur when $\mathbf{f}_1 = \lambda \mathbf{f}_3$ and $\mathbf{f}_2 = \lambda \mathbf{f}_3$, which are analogous to the geometry illustrated by the figure.) The first two columns of det j are iden-

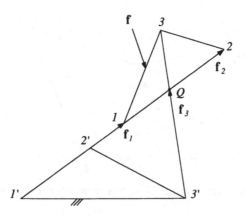

Figure 2.28 Singularity configuration; a pair of connector forces whose lines of action coalesce.

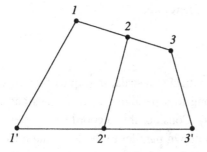

Figure 2.29 In-parallel manipulator with special geometry.

tical to a scalar multiple and hence det $j = 0$. It is clear from Figure 2.28 that there are two independent connector forces $\mathbf{f}_1 + \mathbf{f}_2 = (1 + \lambda)\mathbf{f}_1$ and $\mathbf{f}_3$. The lines of action of these forces meet at a point Q, and their resultant thus lies on the pencil of lines through Q. Any force applied to the top platform can only be equilibrated if its line of action passes through Q.

Finally, consider a parallel manipulator specialized to the extent that all three points of connection of both the moving and fixed platforms are collinear (see Fig. 2.29). Three types of singular configurations occur, which are illustrated by Figure 2.30. It is interesting to note that for this special geometry all three connector forces can become collinear, simultaneously.

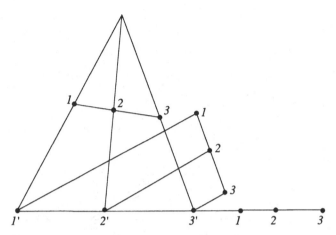

Figure 2.30 Three singularity configurations.

We may conclude that singular or special configurations of in-parallel planar manipulators can be described elegantly by the geometry of pencils of concurrent or parallel lines. Other special cases occur when two or more lines coalesce.

EXERCISE 2.4

Exercise Figure 2.4 illustrates an in-parallel manipulator with special geometry (see also Fig. 2.29) with the moving platform in an initial position for which the coordinates of points 1, 2, and 3 are, respectively, $(0, 3)$, $(0.4\sqrt{2}, 2.4)$, and $(0.95 \sqrt{2}, 1.7)$.

1. Write the unitized coordinates $[c, s, p]^T$ of the force f which passes through point 2 and is parallel to the x axis as shown in the figure.

2. Give the platform self-parallel displacements in increments of 0.1 inches away from the initial position up to 2.9 inches. For each increment compute the coordinates of the points 1, 2, and 3, i.e., $(x_{1n}, 3)$, $(x_{2n}, 2.4)$, and $(x_{3n}, 1.7)$ for $n = 0$, $1, \ldots, 29$.

3. Compute the corresponding connector lengths $(\ell_{1n}, \ell_{2n}, \ell_{3n})$.

4. Use Grassmann's method to compute the corresponding sets of unitized line connector coordinates $[c_{in}, s_{in}, p_{in}]$ for $i = 1, 2, 3$, and $n = 0, 1, \ldots, 29$. Hence, compute the matrix $[j]$ for each position and, correspondingly, compute the inverse $[j]^{-1}$.

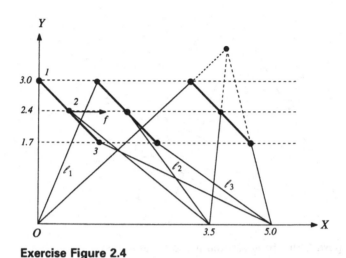

Exercise Figure 2.4

5. Compute the resulting leg forces for each position using

$$\begin{bmatrix} f_1 \\ f_2 \\ f_3 \end{bmatrix}_m = [j]_n^{-1} \begin{bmatrix} c \\ s \\ p \end{bmatrix}.$$

You may assume that $f = 1$ unit of force.

6. Plot the equilibrating force in each of the three connectors against the horizontal position x. Label each curve and state whether a connector is in tension or compression.

2.10 Statically redundant parallel manipulators

Let us consider the static analysis of a platform with four connectors (see Fig. 2.31). Here the force on the top platform and the connector forces are related by

$$\hat{w} = f_1 \begin{bmatrix} c_1 \\ s_1 \\ p_1 \end{bmatrix} + f_2 \begin{bmatrix} c_2 \\ s_2 \\ p_2 \end{bmatrix} + f_3 \begin{bmatrix} c_3 \\ s_3 \\ p_3 \end{bmatrix} + f_4 \begin{bmatrix} c_4 \\ s_4 \\ p_4 \end{bmatrix}. \tag{2.94}$$

Equation (2.94) can be used to compute the resultant of the connector forces. However, the *reverse static analysis is not possible*. It is not possible to com-

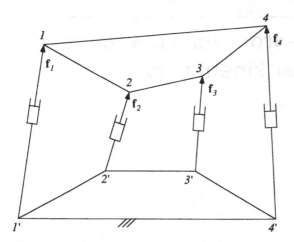

Figure 2.31 A redundant parallel manipulator.

pute a unique set of magnitudes $f_1, f_2, f_3,$ and f_4 which are superabundant by one. Equation 2.94 can be expressed in the matrix form by

$$\hat{w} = j \, \lambda, \qquad (2.95)$$

where

$$j = \begin{bmatrix} c_1 & c_2 & c_3 & c_4 \\ s_1 & s_2 & s_3 & s_4 \\ p_1 & p_2 & p_3 & p_4 \end{bmatrix}, \qquad (2.96)$$

and

$$\lambda = \begin{bmatrix} f_1 \\ f_2 \\ f_3 \\ f_4 \end{bmatrix}. \qquad (2.97)$$

Clearly, j is noninvertible. Systems with four or more connectors are statically indeterminate, and their solution is beyond the scope of this text.

3

First-order instantaneous planar kinematics

3.1 The definition of a rotor

We will consider a rigid lamina that moves on a fixed reference plane and instantaneously rotates about a given fixed axis parallel to a unit vector **k** with an angular speed ω. The angular velocity is given by $\boldsymbol{\omega} = \omega\mathbf{k}$ (see Fig. 3.1). All points on the lamina which do not lie on the axis of rotation are moving on circular paths, and the instantaneous velocity of a point specified by a vector **r** drawn from a point Q in the fixed plane which lies on the axis of rotation is given by

$$\mathbf{v} = \boldsymbol{\omega} \times \boldsymbol{r}. \tag{3.1}$$

In what follows, we choose some reference point O in the fixed plane which is not on the axis of rotation (see Fig. 3.2), and determine the velocity $\mathbf{v}_o$ of a point in the moving lamina instantaneously coincident with O.

The substitutions of $\mathbf{v} = \mathbf{v}_o$ and $\mathbf{r} = -\mathbf{r}_o$ in the left and right sides of (3.1) yield

$$\mathbf{v}_o = \boldsymbol{\omega} \times (-\mathbf{r}_o) = \mathbf{r}_o \times \boldsymbol{\omega}, \tag{3.2}$$

where $\mathbf{v}_o$ is the required velocity of the point on the moving lamina instantaneously coincident with the point O in the frame of reference.

We now define an instantaneous rotation as a rotor, which has the coordinates

$$\hat{\imath} = \{\boldsymbol{\omega}; \mathbf{v}_o\}, \tag{3.3}$$

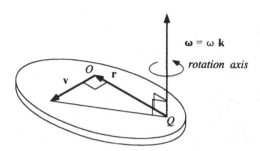

Figure 3.1 Instantaneous rotation.

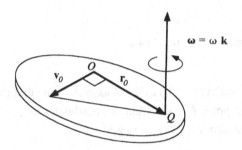

Figure 3.2 Rotation with the reference point at O.

where $\mathbf{v}_o = \mathbf{r}_o \times \boldsymbol{\omega}$.

This is analogous to the representation of the coordinates of a force

$$\hat{w} = \{\mathbf{f}; \mathbf{c}_o\}, \tag{3.4}$$

where $\mathbf{c}_o$ is a couple and $\mathbf{c}_o = \mathbf{r}_o \times \mathbf{f}$.

A comparison of (3.3) and (3.4) shows that the vector forms of a force and a rotor are analogous. The direct analogy between the statics and instantaneous kinematics of a rigid body is becoming apparent. In statics a directed line segment represents the "rectilinear" concept of force, whereas in kinematics a directed line segment represents the "circular" concept of a rotor.

3.2   The coordinates of a line parallel to the Z axis

Any directed line segment parallel to the Z axis can be expressed by

$$\mathbf{S} = N\mathbf{k}, \tag{3.5}$$

where $|\mathbf{S}| = N$. Clearly, $N\mathbf{k}$ is a *free vector* and it can be associated with any

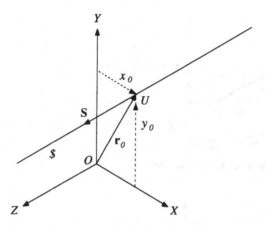

Figure 3.3 A line parallel to the Z axis.

line drawn parallel or antiparallel ($N < 0$) to the Z axis. Assume that the line segment passes through the point U with x and y coordinates (x_o, y_o). The moment of the line segment about the origin is given by

$$\mathbf{r}_o \times \mathbf{S} = (x_o\mathbf{i} + y_o\mathbf{j}) \times N\mathbf{k}$$

$$= (y_oN)\mathbf{i} + (-x_oN)\mathbf{j}, \tag{3.6}$$

which can be expressed in the form

$$\mathbf{r}_o \times \mathbf{S} = P\mathbf{i} + Q\mathbf{i}, \tag{3.7}$$

where $P = y_oN$ and $Q = -x_oN$ are the components of the moment about the X and Y axes, respectively.

The three numbers N, P, and Q are used as the coordinates of the line. They are *homogeneous*: λN, λP, and λQ (where λ is a nonzero scalar) determine the same line. *The units of N, P, and Q are not consistent.* N has dimension $(\text{length})^1$, while P and Q have dimensions of area, namely $(\text{length})^2$. Because of this inconsistency in dimensions, the coordinates are represented by the ordered triple of real numbers $\{N; P, Q\}$ with the semicolon separating N from P and Q. Furthermore, when $N = 1$, $P = y_o$, and $Q = -x_o$. *In other words, the coordinates for a directed line segment of unit length parallel to the Z axis and passing through a point with coordinates (x_o, y_o) are* $\{1; y_o, -x_o\}$. Figure 3.3 shows that the moments of the unit vector $\mathbf{S}$ about the X and Y axes are y_o and $-x_o$, respectively. If the length scales on the axes

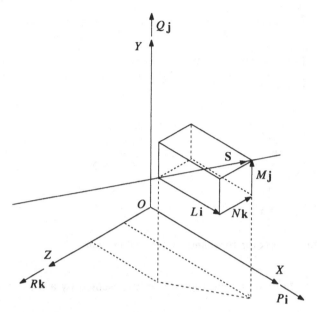

Figure 3.4 A line in three dimensions.

were altered from say, inches to millimeters, then to get the *same* line, P and Q would need to be larger by a factor of 25.4.

The reader may recall that the homogeneous coordinates for a line in the XY plane were expressed by $\{L, M; R\}$ (see Section 2.2). The coordinates (L, M) determined the direction of the line, and R, the moment of a line segment about the Z axis. It has been demonstrated that for a line parallel to the Z axis the direction ratio is N, and the coordinates (P, Q) are moments about the X and Y axes, respectively. These two representations are consistent and they can be expressed as

$\{L, M, 0; 0, 0, R\}$ Line in XY plane,

$\{0, 0, N; P, Q, 0\}$ Line parallel to Z axis (perpendicular to the XY plane).

In fact, the homogeneous coordinates for a line in three-dimensional space are the sextuple $\{L, M, N; P, Q, R\}$ (see Fig. 3.4).

3.3   A useful vector representation of a rotor

Figure 3.5 illustrates a lamina rotating about an axis through point 1 and normal to the page, the XY plane, with an instantaneous angular

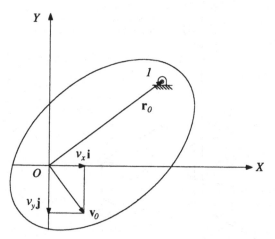

Figure 3.5 A rotor modeled by a revolute pair.

speed ω. The instantaneous motion can be represented by a rotor with co-ordinates

$$\hat{t} = \{\omega; \mathbf{v}_o\}, \tag{eq. 3.3}$$

where $\mathbf{v}_o = \mathbf{r}_o \times \omega$ is the *velocity of a point in the lamina coincident with a reference point O fixed in the reference XY plane*. This representation can be related directly to the line \$ passing through point 1 which has unit coordinates

$$\hat{s} = \{\mathbf{k}; \mathbf{S}_o\}. \tag{3.8}$$

Clearly, $|\mathbf{k}| = 1$, and $\mathbf{S}_o = \mathbf{r}_o \times \mathbf{k} = y_o\mathbf{i} - x_o\mathbf{j}$.

In addition,

$$\omega = \omega\mathbf{k},$$

$$\mathbf{v}_o = (\omega y_o)\mathbf{i} + (-\omega x_o)\mathbf{j} = v_x\mathbf{i} + v_y\mathbf{j}, \tag{3.9}$$

where $v_x = \omega y_o$ and $v_y = -\omega x_o$ are the components of $\mathbf{v}_o$ in the X and Y directions. From (3.3) and (3.9):

$$\hat{t} = \omega \{\mathbf{k}; y_o\mathbf{i} - x_o\mathbf{j}\}. \tag{3.10}$$

The coordinates for the rotor are thus $\{\omega; \omega y_o, -\omega x_o\}$ or $\{\omega; v_x, v_y\}$. The substitution of (3.8) into the right side of (3.10) yields

$$\hat{t} = \omega \hat{s}. \tag{3.11}$$

The rotor is thus expressed as a scalar multiple of the unit line vector $\hat{s}$, defined by (3.8).

3.4   Infinitesimal displacements of a rigid lamina

Assume now that the rigid lamina (see Fig. 3.1) moves on the fixed reference plane by undergoing a small rotation $\delta\phi$ about an axis parallel to the unit vector $\mathbf{k}$ (see Fig. 3.6).

The tangential displacement of a point on the moving lamina coincident with a point O in the frame of reference (see also Fig. 3.6) is given approximately by

$$\delta\mathbf{r}_o = \mathbf{r}_o \times \delta\phi\mathbf{k}. \tag{3.12}$$

The tangential and rotational displacements $\delta\mathbf{r}_o$ and $\delta\phi$ are easy to relate to the tangential and angular velocities $\mathbf{v}_o$ and $\boldsymbol{\omega}$ by dividing both sides of (3.12) by a small time increment δt. As $\delta t \to O$,

$$\frac{\delta\mathbf{r}_o}{\delta t} \to \frac{d\mathbf{r}_o}{dt} = \mathbf{v}_o \quad \text{and} \quad \frac{\delta\phi}{\delta t}\mathbf{k} \to \frac{d\phi}{dt}\mathbf{k} = \omega\mathbf{k} = \boldsymbol{\omega},$$

and (3.12) can be written in the form $\mathbf{v}_o = \mathbf{r}_o \times \boldsymbol{\omega}$, which is precisely (3.2).

Now

$$\mathbf{r}_o = x_o\mathbf{i} + y_o\mathbf{j}, \tag{3.13}$$

and therefore

$$\delta\mathbf{r}_o = (x_o\mathbf{i} + y_o\mathbf{j}) \times \delta\phi\mathbf{k} = \delta\phi y_o\mathbf{i} - \delta\phi x_o\mathbf{j}, \tag{3.14}$$

which can also be expressed in the form

$$\delta\mathbf{r}_o = \delta x_o\mathbf{i} + \delta y_o\mathbf{j}, \tag{3.15}$$

where

$$\delta x_o = \delta\phi y_o, \quad \delta y_o = -\delta\phi x_o \tag{3.16}$$

are the components of $\delta\mathbf{r}_o$ in the X and Y directions.

The infinitesimal displacement will be denoted by

$$\delta\hat{d} = \delta\phi\{\mathbf{k}; y_o\mathbf{i} - x_o\mathbf{j}\}, \tag{3.17}$$

which is completely analogous to the twist

$$\hat{t} = \omega\{\mathbf{k}; y_o\mathbf{i} - x_o\mathbf{j}\}. \tag{eq. 3.10}$$

The coordinates for $\delta\hat{d}$ are thus $\{\delta\phi; \delta\phi y_o, -\delta\phi x_o\}$ or $\{\delta\phi; \delta x_o, \delta y_o\}$.

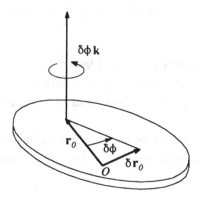

Figure 3.6 Small rotation $\delta\phi$.

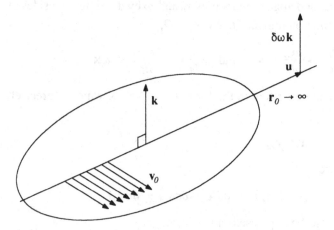

Figure 3.7 A pure translation.

3.5 A representation of pure translation

Assume that the rigid lamina (Fig. 3.7) is moving with a pure translational velocity $\mathbf{v}_o$, that is, every point in the lamina moves with this velocity. Clearly, $\boldsymbol{\omega} = \mathbf{O}$ and the representation of this motion is given by

$$\hat{\mathbf{t}} = \{\mathbf{O}; \mathbf{v}_o\}, \tag{3.18}$$

where $\{\mathbf{O}; \mathbf{v}_o\}$ is a free vector and not a line-bound vector. The instantaneous translation $\mathbf{v}_o$ of the lamina, however, can be considered a rotor with infinitesimal magnitude $\delta\omega$ whose axis $\mathbf{k}$ is perpendicular to the plane of translation. If we write $\mathbf{r}_o = r_o\mathbf{u}$, where $\mathbf{u}$ is a unit vector parallel to $\mathbf{r}_o$, then $\mathbf{v}_o = (\delta\omega r_o)(\mathbf{u} \times \mathbf{k})$ is finite as $\delta\omega \to 0$ and $r_o \to \infty$. This is analogous to the representation of a pure couple (see Section 2.3).

3.6 Ray and axis coordinates of a line and a rotor

Thus far, the coordinates $\hat{s}$ for a line $ have been expressed by the ordered pair of vectors $\{\mathbf{S}; \mathbf{S}_o\}$ and

$$\hat{s} = \{\mathbf{S}; \mathbf{S}_o\}. \tag{3.19}$$

In this representation, the coordinates $\hat{s}$ are defined as ray coordinates because the line that can be drawn through the two points labeled 1 and 2 may be imagined as an actual ray or narrow beam of light originating at a point (see Fig. 3.8).

The coordinates for the same line $ can be expressed equally by the ordered pair of vectors $\{\mathbf{S}_o; \mathbf{S}\}$ and denoted by $\hat{S}$, where

$$\hat{S} = \{\mathbf{S}_o; \mathbf{S}\}. \tag{3.20}$$

In this representation, the coordinates $\hat{S}$ are defined as axis coordinates because the line $ is considered as the meet of two planes (see Fig. 3.8). A pencil of planes could be drawn through $, which can be considered the axis of the pencil. (The lower case $\hat{s}$ and the upper case $\hat{S}$ distinguish the two coordinate representations of the same line $.)

It is popular to express the coordinates for a rotor t in axis coordinates. The ray coordinates for t are

$$\hat{t} = \{\boldsymbol{\omega}; \mathbf{v}_o\}, \tag{eq. 3.3}$$

or

$$\hat{t} = \omega\,\hat{s}, \tag{eq. 3.11}$$

where

$$\hat{s} = \{\mathbf{S}; \mathbf{S}_o\}, \tag{eq. 3.19}$$

and, similarly, the axis coordinates $\hat{T}$ for the same rotor t are

$$\hat{T} = \{\mathbf{v}_o; \boldsymbol{\omega}\}, \tag{3.21}$$

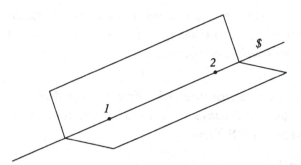

Figure 3.8 A line $ defined by two points or two planes.

or

$$\hat{T} = \omega \, \hat{S}, \tag{3.22}$$

where

$$\hat{S} = \{S_o; S\}. \tag{eq. 3.20}$$

From (3.10) the axis coordinates for a rotor perpendicular to the XY plane are

$$\hat{T} = \omega \, \{y_o\mathbf{i} - x_o\mathbf{j}; \, \mathbf{k}\}. \tag{3.23}$$

The axis coordinates for the rotor are thus $\{\omega y_o, \, -\omega x_o; \, \omega\}$ or $\{v_x, \, v_y; \, \omega\}$, and the axis coordinates for a pure translation are $\{v_x, \, v_y; \, 0\}$.

Analogously, the axis coordinates $\delta\hat{D}$ for an infinitesimal rotation can be written in the form

$$\delta\hat{D} = \delta\phi \, \{y_o\mathbf{i} - x_o\mathbf{j}; \, \mathbf{k}\}. \tag{3.24}$$

Divide the left and right sides of (3.24) by δt and write

$$\hat{T} = \lim_{\delta t \to 0} \frac{\delta\hat{D}}{\delta t} \quad \text{and} \quad \omega = \lim_{\delta t \to 0} \frac{\delta\phi}{\delta t}$$

to yield (3.23). The axis coordinates are thus expressible in the forms $\{\delta\phi y_o, \, -\delta\phi x_o; \, \delta\phi\}$ or $\{\delta x_o, \, \delta y_o; \, \delta\phi\}$ and the coordinates for an infinitesimal translation are $\{\delta x_o, \, \delta y_o; \, 0\}$.

From here on, the coordinates for lines and rotors perpendicular to the XY plane will be expressed in axis coordinates.

3.7   Translation and rotation of coordinate systems

The Plücker coordinates of a unit line bound vector perpendicular to the XY plane though point U with coordinates (x, y) are $\{0, y, -x; 0, 0, 1\}$, which will be written in the abbreviated form

$$\{y, -x; 1\}. \tag{3.25}$$

The coordinates after the translation are

$$\{y', -x'; 1\}. \tag{3.26}$$

The translation is illustrated by Figure 3.9 and

$$x = x' + a, \quad y = v' + b. \tag{3.27}$$

It is clear that point U does not change under this transformation. Therefore, neither does the directed line segment pointing out of the page through the point U nor the line of unlimited length passing through U. All these geometric elements, point, directed line segment, and line, are therefore invariant with a translation of coordinate system, although the actual coordinates in the translated system are different. The arrangement of (3.27) in the form of (3.25) yields

$$y = y' + b, \quad -x = -x' - a, \quad 1 = 1, \tag{3.28}$$

which can be expressed in matrix form as

$$\begin{bmatrix} y \\ -x \\ 1 \end{bmatrix} = \begin{bmatrix} 1 & 0 & b \\ 0 & 1 & -a \\ 0 & 0 & 1 \end{bmatrix} \begin{bmatrix} y' \\ -x' \\ 1 \end{bmatrix}. \tag{3.29}$$

The coordinate $\{0, 0; 1\}$ remains unchanged under the translation and it is therefore independent of the coordinate system. It defines a *free vector* **k** parallel to the Z axis and its value remains unchanged under parallel displacement. All three coordinates $\{y, -x; 1\}$ define a line bound vector and the coordinates $\{y', -x'; 1\}$ define precisely the identical line bound vector, the line segment (or rotor) in the second coordinate system. As stated before, the coordinates $\{y, -x; 0\}$ for the moment change are origin dependent. However, since y' and $-x'$ are solely functions of b and a, they quantify the moment about a different origin O'.

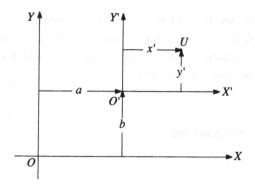

Figure 3.9 A translation of a coordinate system.

The rotation of a rectangular coordinate system through an angle ϕ is illustrated in Figure 3.10. From (2.42),

$$x' = x''\cos \phi - y'' \sin \phi,$$
$$y' = x''\sin \phi + y'' \cos \phi. \tag{3.30}$$

Therefore

$$y' = y'' \cos \phi - (-x'') \sin \phi,$$
$$-x' = y'' \sin \phi + (-x'') \cos \phi,$$
$$1 = 1, \tag{3.31}$$

which can be expressed in the matrix form

$$\begin{bmatrix} y' \\ -x' \\ 1 \end{bmatrix} = \begin{bmatrix} \cos \phi & -\sin \phi & 0 \\ \sin \phi & \cos \phi & 0 \\ 0 & 0 & 1 \end{bmatrix} \begin{bmatrix} y'' \\ -x'' \\ 1 \end{bmatrix}. \tag{3.32}$$

The matrix equation (3.32) expresses the coordinates of the rotor in the old coordinate system in terms of the coordinates of the rotor in the new coordinate system. For this transformation the coordinate {0, 0; 1} remains unchanged. The coordinates {y'', $-x''$; 0} in the new coordinate system depend solely upon the original coordinates {y', $-x'$; 0}, and the relationship of y'', $-x''$ to y', $-x'$ is the same as (x'', y'') to (x', y').

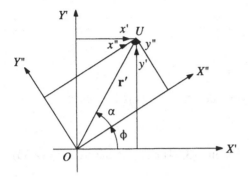

Figure 3.10 A rotation of a coordinate system.

3.8 Induced line (rotor) transformation of the Euclidean group

For both translation and rotation of the coordinate system, the induced line transformation obtained from (3.29) and (3.32) can be expressed in the form

$$\begin{bmatrix} y \\ -x \\ 1 \end{bmatrix} = [E] \begin{bmatrix} y'' \\ -x'' \\ 1 \end{bmatrix}, \tag{3.33}$$

where

$$[E] = \begin{bmatrix} 1 & 0 & b \\ 0 & 1 & -a \\ 0 & 0 & 1 \end{bmatrix} \begin{bmatrix} c & -s & 0 \\ s & c & 0 \\ 0 & 0 & 1 \end{bmatrix} = \begin{bmatrix} c & -s & b \\ s & c & -a \\ 0 & 0 & 1 \end{bmatrix}. \tag{3.34}$$

3.9 Relationship between [e] and [E]

In Chapter 2 we derived and expressed the transformation [e] for a force or line in the *XY* plane under the Euclidean group in the form

$$[e] = \begin{bmatrix} c & -s & 0 \\ s & c & 0 \\ as - bc & ac + bs & 1 \end{bmatrix}. \tag{eq. 2.53}$$

Forming the inverse of $[e]$ yields

$$[e]^{-1} = \begin{bmatrix} c & -s & b \\ s & c & -a \\ 0 & 0 & 1 \end{bmatrix}^T . \tag{3.35}$$

Clearly, from (3.34) the right side of (3.35) is $[E]^T$. Hence

$$[e]^{-1} = [E]^T. \tag{3.36}$$

Analogously, forming $[E]^{-1}$ using (3.34) and comparing with (2.53) yields

$$[e]^T = [E]^{-1}. \tag{3.37}$$

──────── **EXERCISE 3.1** ────────────────────

1. Determine the coordinates of the points of intersection of the following lines with the $Z = 0$ plane: $\{12, 16; 4\}$, $\{12, -16; 4\}$, $\{-12, 16; 4\}$, $\{12, 16; -4\}$, and $\{-12, -16; -4\}$.
2. Determine the 3×3 matrix representations $[A]$ and $[B]$ of the induced line transformation for a translation and rotation of an XY coordinate system. Obtain $[A]$ and $[B]$ for $a = 2$ inches, $b = 3$ inches, and $\phi = 60$ degrees. Finally, determine the product $[A][B]$.

3.10 The first-order instantaneous kinematics of a 3R serial manipulator

The end link a_{34} that carries the end effector is connected to the ground via three serially connected revolute joints. Assume at the outset that the geometric configuration (θ_1, θ_2, θ_3) is known (see Fig. 3.11).

Assume that three rotors with magnitudes ω_1, ω_2, and ω_3 are generated in each joint about the lines $\$_1$, $\$_2$, and $\$_3$, which are parallel to the Z axis. It is important to recognize that ω_1 is the angular velocity of link a_{12} with respect to the ground, ω_2 is the angular velocity of link a_{23} relative to link a_{12}, and ω_3 is the angular velocity of link a_{34} with respect to link a_{23}. The end effector will move on some rotor T whose coordinates are given by

$$\hat{T} = \hat{T}_1 + \hat{T}_2 + \hat{T}_3, \tag{3.38}$$

which can be expressed in the alternative form

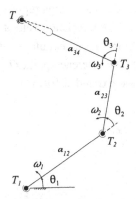

Figure 3.11 A 3R serial manipulator.

$$\hat{T} = \begin{bmatrix} \mathbf{v}_o \\ \boldsymbol{\omega} \end{bmatrix} = \begin{bmatrix} \mathbf{v}_{o1} \\ \omega_1 \end{bmatrix} + \begin{bmatrix} \mathbf{v}_{o2} \\ \omega_2 \end{bmatrix} + \begin{bmatrix} \mathbf{v}_{o3} \\ \omega_3 \end{bmatrix}, \tag{3.39}$$

$$\hat{T} = \begin{bmatrix} \mathbf{v}_o \\ \boldsymbol{\omega} \end{bmatrix} = \omega_1 \hat{S}_1 + \omega_2 \hat{S}_2 + \omega_3 \hat{S}_3$$

$$= \omega_1 \begin{bmatrix} \mathbf{S}_{o1} \\ \mathbf{k} \end{bmatrix} + \omega_2 \begin{bmatrix} \mathbf{S}_{o2} \\ \mathbf{k} \end{bmatrix} + \omega_3 \begin{bmatrix} \mathbf{S}_{o3} \\ \mathbf{k} \end{bmatrix}. \tag{3.40}$$

Therefore, from (3.40):

$$\boldsymbol{\omega} = \omega_1 \mathbf{k} + \omega_2 \mathbf{k} + \omega_3 \mathbf{k}, \tag{3.41}$$

and

$$\mathbf{v}_o = \omega_1 \mathbf{S}_{o1} + \omega_2 \mathbf{S}_{o2} + \omega_3 \mathbf{S}_{o3}. \tag{3.42}$$

From (3.41):

$$\omega = |\boldsymbol{\omega}| = \omega_1 + \omega_2 + \omega_3. \tag{3.43}$$

From (3.42):

$$\mathbf{S}_o = \frac{\omega_1}{\omega} \mathbf{S}_{o1} + \frac{\omega_2}{\omega} \mathbf{S}_{o2} + \frac{\omega_3}{\omega} \mathbf{S}_{o3}. \tag{3.44}$$

In the previous equations $\hat{T}_1$, $\hat{T}_2$, and $\hat{T}_3$ represent, respectively, the coordinates of the motions of link a_{12} with respect to the ground, link a_{23} with respect to link a_{12}, and link a_{34} with respect to a_{23}. The resultant $\hat{T}$ thus represents the coordinates of link a_{34} with respect to the ground. The translation velocities v_{o1}, v_{o2}, and v_{o3}, are, respectively, the velocities of points in the laminas containing links a_{12}, a_{23}, and a_{34} which are coincident with the reference point O.

It is often convenient to express (3.40) in the expanded form

$$\hat{T} = \omega_1 \begin{bmatrix} y_1 \\ -x_1 \\ 1 \end{bmatrix} + \omega_2 \begin{bmatrix} y_2 \\ -x_2 \\ 1 \end{bmatrix} + \omega_3 \begin{bmatrix} y_3 \\ -x_3 \\ 1 \end{bmatrix}, \tag{3.45}$$

where

$$\begin{bmatrix} y_i \\ -x_i \\ 1 \end{bmatrix}, \quad (i = 1, 2, 3)$$

are the Plücker coordinates of the lines that pass through the joint axes.

Equation 3.45 can be expressed in matrix form by

$$\hat{T} = J\gamma, \tag{3.46}$$

where

$$J = \begin{bmatrix} y_1 & y_2 & y_3 \\ -x_1 & -x_2 & -x_3 \\ 1 & 1 & 1 \end{bmatrix} \tag{3.47}$$

is a 3×3 matrix, and

$$\hat{T} = \begin{bmatrix} v_{ox} \\ v_{oy} \\ \omega \end{bmatrix}, \quad \gamma = \begin{bmatrix} \omega_1 \\ \omega_2 \\ \omega_3 \end{bmatrix} \tag{3.48}$$

are 3×1 column vectors.

When the magnitudes ω_1, ω_2, and ω_3 are specified it is a simple matter to compute the coordinates $\hat{T}$ of the resultant rotor using (3.40). This is called the *forward kinematic analysis*. Conversely, when the end effector moves on some rotor with coordinates $\hat{T}$, we need to determine the magnitudes ω_1, ω_2, and ω_3. The solution is easy to obtain from (3.46) and

$$\gamma = J^{-1}\hat{T}, \tag{3.49}$$

where J^{-1} is the inverse of J. This is called the *reverse velocity analysis*. Finally, substituting

$$\hat{T} = \lim_{\delta t \to 0} \frac{\delta \hat{D}}{\delta t} \quad \text{and} \quad \omega_i = \lim_{\delta t \to 0} \frac{\delta \theta_i}{\delta t}$$

into (3.46) and canceling throughout by δt yields

$$\delta \hat{D} = J \delta \theta, \tag{3.50}$$

where

$$\delta \hat{D} = \begin{bmatrix} \delta x_o \\ \delta y_o \\ \delta \phi \end{bmatrix}, \ \delta \theta = \begin{bmatrix} \delta \theta_1 \\ \delta \theta_2 \\ \delta \theta_3 \end{bmatrix}, \tag{3.51}$$

which are analogous to (3.46) and (3.48). Equation 3.50 thus relates the infinitesimal joint displacements to the infinitesimal motion of the end effector. Therefore,

$$\delta \theta = J^{-1} \delta \hat{D}, \tag{3.52}$$

which is analogous to (3.49).

3.11 The geometrical meaning of J^{-1}

Now from (3.47):

$$J^{-1} = \begin{bmatrix} x_3 - x_2 & x_1 - x_3 & x_2 - x_1 \\ y_3 - y_2 & y_1 - y_3 & y_2 - y_1 \\ x_2 y_3 - x_3 y_2 & x_3 y_1 - x_1 y_3 & x_1 y_2 - x_2 y_1 \end{bmatrix}^T / \det J$$

$$= \begin{bmatrix} a_{23} & \hat{s}_{23}^T \\ a_{31} & \hat{s}_{31}^T \\ a_{12} & \hat{s}_{12}^T \end{bmatrix} / \det J, \tag{3.53}$$

where by Grassmann (see Section 2.1) $\hat{s}_{23}$, $\hat{s}_{31}$, and $\hat{s}_{12}$ are the column vectors formed by the unitized coordinates for the lines joining the points 2–3, 3–1, and 1–2 (see Fig. 3.12). For instance, employing the usual notation $\hat{s}_{23}^T = [c_2, s_2; p_2]$ and $[(x_3 - x_2), (y_3 - y_2); (x_2 y_3 - x_3 y_2)] = a_{23} \hat{s}_{23}$, where $a_{23} = \{(x_3 - x_2)^2 + (y_3 - y_2)^2\}^{1/2}$ is the distance between joints 2 and 3. Further-

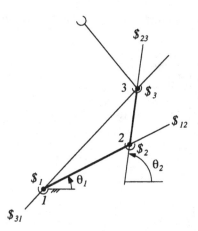

Figure 3.12 The lines associated with a 3*R* serial manipulator.

more, det J can be expressed in three ways. Expansion from the first column of the right side of (3.47) yields

$$\det J = y_1(x_3 - x_2) - x_1(y_3 - y_2) + 1\,(x_2y_3 - x_3y_2), \qquad (3.54)$$

which can be expressed in the form

$$\det J = \hat{S}_1^T a_{23}\hat{s}_{23} = a_{23}\hat{s}_{23}^T\,\hat{S}_1. \qquad (3.55)$$

Analogously, expansion from the second and third columns of the right side of (3.47) yields

$$\det J = \hat{S}_2^T a_{31}\hat{s}_{31} = a_{31}\hat{s}_{31}^T\,\hat{S}_2, \qquad (3.56)$$

and

$$\det J = \hat{S}_3^T a_{12}\hat{s}_{12} = a_{12}\hat{s}_{12}^T\,\hat{S}_3. \qquad (3.57)$$

The substitution of (3.53) into (3.49) yields

$$\gamma = \begin{bmatrix} \omega_1 \\ \omega_2 \\ \omega_3 \end{bmatrix} = \left(\begin{bmatrix} a_{23}\hat{s}_{23}^T \\ a_{31}\hat{s}_{31}^T \\ a_{12}\hat{s}_{12}^T \end{bmatrix} / \det J \right) \hat{T}. \qquad (3.58)$$

Therefore, from (3.58) and (3.55)–(3.57):

$$\omega_1 = \hat{s}_{23}^T\,\hat{T} / \hat{s}_{23}^T\hat{S}_1, \qquad (3.59)$$

$$\omega_2 = \hat{s}_{31}^T \, \hat{T} / \hat{s}_{31}^T \hat{S}_2, \tag{3.60}$$

$$\omega_3 = \hat{s}_{12}^T \, \hat{T} / \hat{s}_{12}^T \hat{S}_3. \tag{3.61}$$

Chapter 4 shows that (3.59)–(3.61) can be obtained directly from (3.40) by forming so-called reciprocal products.

─────── **EXERCISE 3.2** ───────────────────────

1.   (a) A lamina is instantaneously rotating about a point C with coordinates (x_c, y_c). The velocity v_p of a point P is known. The coordinates of P are (x_p, y_p). Obtain an expression for the angular velocity ω in terms of v_p and the coordinates of C and P. Hence, derive an expression for the velocity v_o of a point in the lamina coincident with the reference point O and draw its direction on Exercise Figure 3.2(1a).

(b) A lamina is instantaneously rotating about a point C with coordinates (x_c, y_c), where $x_c = 0$ and $y_c = 2$ inches. The velocity v_p of a point P is $v_p = 1$ in./sec. (see Exr. Fig. 3.2(1b)). The coordinates of P are (x_p, y_p), where $x_p = 1$ inch and $y_p = 2$ inch. Use the expression from part (a) to determine ω and the velocity v_o of a point in the lamina coincident with the reference point O. Compute the joint velocities ω_1, ω_2, and ω_3 for the following 3R manipulator. (Note: For a single calculation it is as easy to solve three simultaneous equations for ω_1, ω_2, and ω_3 as it is to invert a 3×3 matrix.)

2. Compute the joint velocities ω_1, v_2, and ω_3 of the RPR manipulator for each end-effector velocity (see Exr. Fig. 3.2(2)):

$$\hat{T} = \omega \begin{bmatrix} 0 \\ 0 \\ 1 \end{bmatrix}, \quad \hat{T} = v_x \begin{bmatrix} 1 \\ 0 \\ 0 \end{bmatrix}, \quad \hat{T} = v_y \begin{bmatrix} 0 \\ 1 \\ 0 \end{bmatrix}.$$

Write a short comment on each result; explain how it makes sense physically.

3.   (a) Write the coordinates $\hat{T}$ for the instant motion of the end effector of the 3R robot analyzed in 1(b).

(b) Use Grassmann (see equation (2.9)) to determine the coordinates of $\hat{s}_{23}$, $\hat{s}_{31}$, and $\hat{s}_{12}$ of the lines $\$_{23}$, $\$_{31}$, and $\$_{12}$ for the 3R robot shown in Exercise Figure 3.2(3).

(c) Write the coordinates $\hat{S}_1$, $\hat{S}_2$, and $\hat{S}_3$ of the lines $\$_1$, $\$_2$, and $\$_3$ through the points $(0, 0)$, $(3, 0)$, and $(2, 2)$ which are perpendicular to the XY plane.

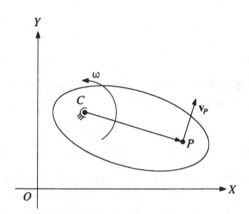

Exercise Figure 3.2(1a)

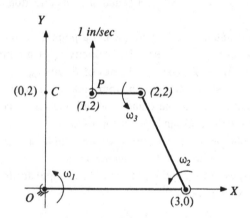

Exercise Figure 3.2(1b)

(d) Show that $\hat{s}_{23}^T \hat{S}_2 = \hat{s}_{23} \hat{S}_3 = 0$, $\hat{s}_{31}^T \hat{S}_3 = \hat{s}_{31}^T \hat{S}_1 = 0$, and $\hat{s}_{12}^T \hat{S}_1 = \hat{s}_{12}^T \hat{S}_2 = 0$.

(e) Compute ω_1, ω_2, and ω_3 using (3.59)–(3.61).

4. It has been shown that for a 3R manipulator: det $J = a_{12}\hat{s}_{12}^T\hat{S}_3$ (see equation (3.57)) which is twice the area of the triangle 123 (see also Exr. Fig. 3.2(4)).

(a) Deduce that det $J = a_{12}a_{23} \sin \theta_2$.

(b) Deduce that the maximum value for det J is det $J_{max} = a_{12}a_{23}$ and that $\lambda = $ det $J/$det $J_{max} = \sin \theta_2$.

(c) Sketch the graph $\lambda = \sin \theta_2$ for $0 \le \theta_2 \le 360$ degrees. Draw the pairs configurations of the manipulator for $\lambda = 0$ and $\lambda = \pm 1$.

Note: λ is independent of the manipulator dimensions.

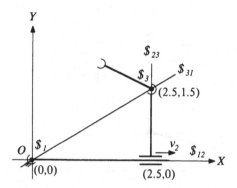

Exercise Figure 3.2(2)

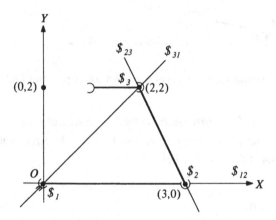

Exercise Figure 3.2(3)

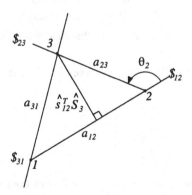

Exercise Figure 3.2(4)

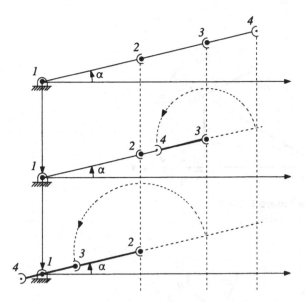

Figure 3.13 Singularity configurations of a 3R serial manipulator.

3.12 Singularity configurations of a serial manipulator

Choose the origin of a coordinate system to lie on the first joint axis, $x_1 = y_1 = 0$. From (3.47) det $J = 0$ when

$$\frac{y_2}{x_2} = \frac{y_3}{x_3} = \tan \alpha, \tag{3.62}$$

as illustrated by Figure 3.13. Points 2 and 3 lie on a line through the origin which is defined as an extreme distance line. When a reference point Q in the end effector lies on this line it is at an extreme distance from the first grounded joint.

Rotating the coordinate system so that the rotor axes lie on the X axis yields (see also Fig. 3.14)

$$\hat{T} = \omega \begin{bmatrix} y \\ -x \\ 1 \end{bmatrix} = \begin{bmatrix} 0 & 0 & 0 \\ 0 & -x_2 & -x_3 \\ 1 & 1 & 1 \end{bmatrix} \begin{bmatrix} \omega_1 \\ \omega_2 \\ \omega_3 \end{bmatrix}. \tag{3.63}$$

It is not possible for the end effector to move on a rotor T with coordinates

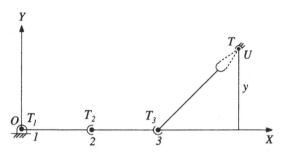

Figure 3.14 Point U must be on X axis.

$\omega\{y, -x; 1\}$ whose axis passes through the point U, no matter what the intensities ω_1, ω_2, and ω_3 of the three rotors T_1, T_2, and T_3 are. Clearly, from (3.63) $\omega y = 0$, and since $y \neq 0$, then $\omega = 0$.

The singularity configurations of $2R$-P and R-$2P$ manipulators are discussed in question 2 of Exercise 3.3.

EXERCISE 3.3

1.   **(a)** Use parallel projection to obtain expressions for the coordinates (x_2, y_2), (x_3, y_3) of the points 2 and 3 of the $2R$ planar manipulator (see Exr. Fig. 3.3(1)).

    (b) Perform a complete reverse displacement analysis, i.e., obtain expressions for $\cos \theta_2$. See also subsection 1.4.1 ($\cos \theta_1$, $\sin \theta_1$).

2. The end effector of the $3R$ manipulator is moving with a pure translational constant velocity v parallel to the x axis, as illustrated in Exercise Figure 3.3(2). Give the end-effector self-parallel displacements in increments $\delta = 0.25$ ($0 \leq \delta \leq 2.25$) and $\delta = 0.025$ ($2.25 < \delta \leq 2.5$) of 0.2 feet away from the initial position $(x_3, y_3) = (0, 2.5)$ ($a_{12} = 2$ ft., $a_{23} = 1.5$ ft.).

    (a) For each set of coordinates (x_3, y_3) compute corresponding sets of θ_2 and θ_1 using the reverse analysis in question 1. Then compute the corresponding values for the coordinates (x_2, y_2).

    (b) Compute the matrix

$$J = \begin{bmatrix} 0 & y_2 & y_3 \\ 0 & -x_2 & -x_3 \\ 1 & 1 & 1 \end{bmatrix}$$

for each increment and correspondingly compute J^{-1}.

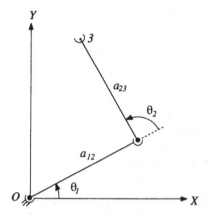

Exercise Figure 3.3(1)

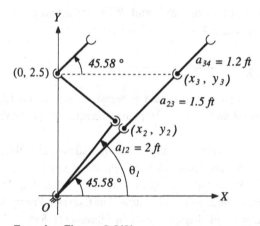

Exercise Figure 3.3(2)

(c) Compute the corresponding angular velocities of the joints

$$\begin{bmatrix} \omega_1 \\ \omega_2 \\ \omega_3 \end{bmatrix} = J^{-1}\hat{T},$$

using a value of $v = 1$ in./sec.

(d) Plot ω_1, ω_2, and ω_3 against x.

3. (a) Write the matrix J for the *PRR* manipulator (see Exr. Fig. 3.3(3a)).

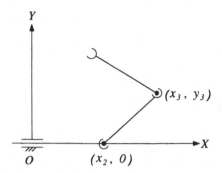

Exercise Figure 3.3(3a)

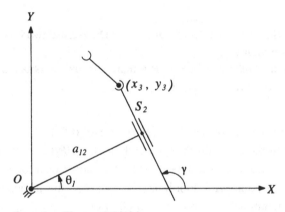

Exercise Figure 3.3(3b)

Show that $\det J = 0$ when $x_2 = x_3$. Draw this configuration and comment on your result.

(b) Write the matrix J for the *RPR* manipulator shown in Exercise Figure 3.3(3b). Verify that

$$x_3 = a_{12} \cos \theta_1 + S_2 \cos \gamma,$$
$$y_3 = a_{12} \sin \theta_1 + S_2 \sin \gamma,$$
$$\gamma = \theta_1 + \pi/2.$$

Use these expressions to show that $\det J = 0$ when $S_2 = 0$. Draw this configuration and comment on your results.

(c) Write the matrix J for the *RRP* manipulator shown in Exercise Figure 3.3(3c). Verify the expressions $x_2 = a_{12} \cos \theta_1$, $y_2 = a_{12} \sin \theta_1$, $\gamma = \theta_1 +$

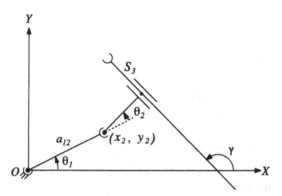

Exercise Figure 3.3(3c)

$\theta_2 + \pi/2$, and show that det $J = 0$ when $\sin \theta_2 = 0$. Draw these two configurations and comment on your results.

(d) Write the J matrices for the 2P-R manipulators and obtain the singularity condition for the PRP manipulator.

3.13 Kinematics of redundant serial manipulators

A serial manipulator with four joints, that is, a *PRRR* manipulator, is shown in Figure 3.15. Assume that the end effector moves on a rotor whose axis passes through a point with coordinates (x, y); then the instantaneous kinematics of the manipulator requires the solution of

$$\hat{T} = \omega \begin{bmatrix} y \\ -x \\ 1 \end{bmatrix} = v_1 \begin{bmatrix} 1 \\ 0 \\ 0 \end{bmatrix} + \omega_2 \begin{bmatrix} 0 \\ -x_2 \\ 1 \end{bmatrix} + \omega_3 \begin{bmatrix} y_3 \\ -x_3 \\ 1 \end{bmatrix} + \omega_4 \begin{bmatrix} y_4 \\ -x_4 \\ 1 \end{bmatrix}. \quad (3.64)$$

Clearly, (3.64) can be used to compute the resultant rotor for a given set of joint speeds. However the *inverse velocity problem does not have a unique solution*: It is not possible, in general, to solve uniquely for $\{v_1, \omega_2, \omega_3, \omega_4\}$ because the matrix

$$J = \begin{bmatrix} 1 & 0 & y_3 & y_4 \\ 0 & -x_2 & -x_3 & -x_4 \\ 0 & 1 & 1 & 1 \end{bmatrix} \quad (3.65)$$

is noninvertible. The inverse velocity analysis of serial manipulators with four or more rotors is indeterminate.

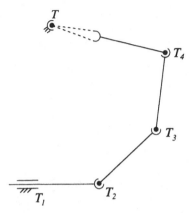

Figure 3.15 A *PRRR* redundant manipulator.

3.14 First-order instantaneous kinematics of a closed-loop 4R mechanism

The twist equation for the closed-loop 4R mechanism (see Fig. 3.16) can be expressed as

$$\hat{T}_4 + \hat{T}_3 + \hat{T}_2 + \hat{T}_1 = 0, \qquad (3.66)$$

where $\hat{T}_4$, $\hat{T}_1$, $\hat{T}_2$, and $\hat{T}_3$ are the coordinates of the rotors through points 4, 1, 2, and 3 which have the coordinates $(0, 0)$, $(x_1, 0)$, (x_2, y_2), and (x_3, y_3), respectively. Therefore

$$\omega_4 \begin{bmatrix} 0 \\ 0 \\ 1 \end{bmatrix} + \omega_1 \begin{bmatrix} 0 \\ -x_1 \\ 1 \end{bmatrix} + \omega_2 \begin{bmatrix} y_2 \\ -x_2 \\ 1 \end{bmatrix} + \omega_3 \begin{bmatrix} y_3 \\ -x_3 \\ 1 \end{bmatrix} = \begin{bmatrix} 0 \\ 0 \\ 0 \end{bmatrix}. \qquad (3.67)$$

Note that the coordinates $\hat{T}_1$, $\hat{T}_2$, $\hat{T}_3$, and $\hat{T}_4$ represent, respectively, the relative motions of link a_{12} with respect to the frame a_{41}, a_{23} with respect to a_{12}, a_{34} with respect to a_{23}, and the relative motion of the *frame a_{41} with respect to a_{34}*. In other words, ω_1, ω_2, and ω_3 are, respectively, the angular velocities of link a_{12} with respect to the frame a_{41}, a_{23} with respect to a_{12}, and a_{34} with respect to a_{23}, *whereas ω_4 is the angular velocity of the frame a_{41} measured relative to a_{34}*. Usually, the input angular velocity ω_4' of the moving

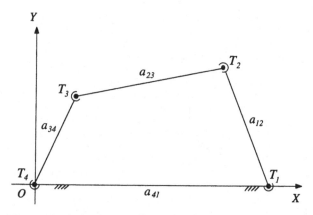

Figure 3.16 A planar 4R mechanism.

link a_{34} with respect to the frame a_{41} is specified. Clearly, in (3.67) and in the following matrix form of (3.67) $\omega_4 = -\omega'_4$.

$$
\omega_4 \begin{bmatrix} 0 \\ 0 \\ 1 \end{bmatrix} = - \begin{bmatrix} 0 & y_2 & y_3 \\ -x_1 & -x_2 & -x_3 \\ 1 & 1 & 1 \end{bmatrix} \begin{bmatrix} \omega_1 \\ \omega_2 \\ \omega_3 \end{bmatrix}. \tag{3.68}
$$

This equation can be used to compute the rotor speeds ω_1, ω_2, and ω_3 for any specified input rotor speed ω'_4 of link a_{34} with respect to a_{41}, with $\omega_4 = -\omega'_4$ provided that

$$
\det J = \begin{vmatrix} 0 & y_2 & y_3 \\ -x_1 & -x_2 & -x_3 \\ 1 & 1 & 1 \end{vmatrix} \neq 0. \tag{3.69}
$$

The expansion of (3.69) from the first row shows that $\det J = 0$ when

$$
\frac{y_2}{x_1 - x_2} = \frac{y_3}{x_1 - x_3} \quad (= \tan \alpha), \tag{3.70}
$$

for which the input a_{34} is in a stationary position relative to the reference frame, and $\omega_4 = 0$, as illustrated by Figure 3.17.

It is interesting to note that although the 4R mechanism is in a limit configuration and it has degenerated into a triangle, namely a structure, the axes of the three joints 1, 2, and 3 are in the same plane perpendicular to the *XY*

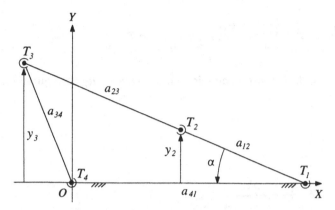

Figure 3.17 Stationary configuration of the input crank.

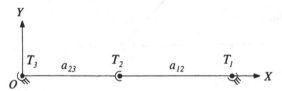

Figure 3.18 An apparent structure which has infinitesimal mobility.

plane, and hence, they are linearly dependent. For convenience, a new *XY* coordinate system is chosen, as illustrated in Figure 3.18.

Since $\omega_4 = 0$, then in this new coordinate system

$$\omega_3 \begin{bmatrix} 0 \\ 1 \end{bmatrix} + \omega_2 \begin{bmatrix} -x_2 \\ 1 \end{bmatrix} + \omega_1 \begin{bmatrix} -x_1 \\ 1 \end{bmatrix} = \begin{bmatrix} 0 \\ 0 \end{bmatrix}. \qquad (3.71)$$

Equation 3.71 is obtained by substituting $y_2 = x_3 = y_3 = 0$ in (3.67). All three rotors span a two-space and belong to a two-system. For example, if one specifies the angular velocity ω_3, then the angular velocities ω_1 and ω_2 can be computed from (3.71). In other words, this apparent structure has instant mobility. This infinitesimal mobility can be detected by pushing up or down on the assemblage.

Figure 3.19 illustrates a 4R mechanism with $a_{41} + a_{12} = a_{34} + a_{23}$ in a folded or an uncertainty configuration. This is because for any input angular speed ω_4, the output crank a_{12} can move in a clockwise or anticlockwise di-

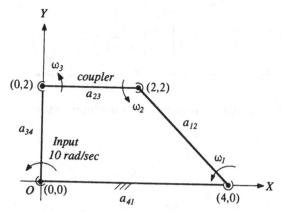

Figure 3.19 A 4*R* mechanism in an uncertainty configuration.

Exercise Figure 3.4(1a)

rection about T_1 relative to the reference frame. All four rotors now belong to a two-system, and

$$\omega_4 \begin{bmatrix} -x_4 \\ 1 \end{bmatrix} + \omega_3 \begin{bmatrix} -x_3 \\ 1 \end{bmatrix} + \omega_2 \begin{bmatrix} -x_2 \\ 1 \end{bmatrix} + \omega_1 \begin{bmatrix} -x_1 \\ 1 \end{bmatrix} + 0. \qquad (3.72)$$

At this configuration, the system is redundant, for if the input velocity is specified it is not possible to determine ω_3, ω_2, and ω_1. In addition, (3.72) can be expressed in the form

$$\omega_4 \begin{bmatrix} -x_4 \\ 1 \end{bmatrix} = - \begin{bmatrix} -x_3 & -x_2 & -x_1 \\ 1 & 1 & 1 \end{bmatrix} \begin{bmatrix} \omega_3 \\ \omega_2 \\ \omega_1 \end{bmatrix}. \qquad (3.73)$$

The solution of (3.73) is not unique. It is not possible to solve uniquely for ω_3, ω_2, and ω_1 because the 2×3 matrix is noninvertible.

——— **EXERCISE 3.4** ———

1. The input crank of the 4*R* mechanism is undergoing an anticlockwise rotation of 10 rads/sec (see Exr. Fig. 3.4(1a)).

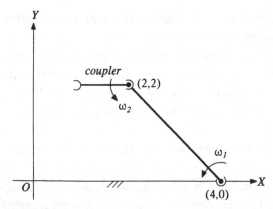

Exercise Figure 3.4(1b)

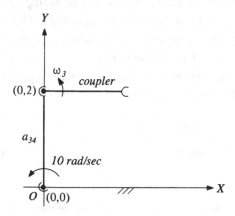

Exercise Figure 3.4(1c)

(a) Compute the angular velocities ω_1, ω_2, and ω_3. (Again it is easier to solve three simultaneous equations for ω_1, ω_2, and ω_3.)

(b) Compute the instant motion of the coupler (i.e., the coordinates for the instant center of rotation and its angular velocity) by disconnecting the mechanism at the third joint and by considering the coupler to be the end effector of the 2R manipulator (see Exr. Fig. 3.4(1b)).

(c) Repeat part b by disconnecting the mechanism at the second joint (see Exr. Fig. 3.4(1c)).

(d) Compare the results obtained in parts (b) and (c) and draw the instant center on your figures.

3.15   Instantaneous kinematics of a 3*R* serial manipulator using displacement equations

We study the instantaneous kinematics of serial manipulators by introducing the concepts of rotors (instantaneous rotations) and instantaneous translations. This method introduces lines and line segments at the outset, establishes a geometrical meaning, and, hence, a firm geometrical foundation for the development.

An alternative and more cumbersome approach that is followed in many texts on robotics is to form time derivatives of the displacement equations for the three degrees of freedom serial manipulators derived in Chapter 1. This method is included for the sake of completeness. It is not recommended by the author, and the reader will undoubtably appreciate the complexity and lack of geometric meaning of the development. By way of example, the instantaneous kinematics for the 3*R* serial manipulator illustrated in Figure 3.20 will be determined by taking time derivatives of the displacement equations. These equations (1.28–1.30) were derived previously and will simply be restated here; the coordinates for a point Q in the end effector together with the orientation of the end effector are given by

$$x_Q = a_{12}c_1 + a_{23}c_{1+2} + a_{34}c_{1+2+3}, \tag{3.74}$$

$$y_Q = a_{12}s_1 + a_{23}s_{1+2} + a_{34}s_{1+2+3}, \tag{3.75}$$

$$\gamma = \theta_1 + \theta_2 + \theta_3. \tag{3.76}$$

The first derivatives of (3.74)–(3.76) can be expressed in the form

$$v_{Qx} \ (= \dot{x}_Q) = \frac{\partial x_Q}{\partial \theta_1}\omega_1 + \frac{\partial x_Q}{\partial \theta_2}\omega_2 + \frac{\partial x_Q}{\partial \theta_3}\omega_3, \tag{3.77}$$

$$v_{Qy} \ (= \dot{y}_Q) = \frac{\partial y_Q}{\partial \theta_1}\omega_1 + \frac{\partial y_Q}{\partial \theta_2}\omega_2 + \frac{\partial y_Q}{\partial \theta_3}\omega_3, \tag{3.78}$$

$$\omega \ (= \dot{\gamma}) = \frac{\partial \gamma}{\partial \theta_1}\omega_1 + \frac{\partial \gamma}{\partial \theta_2}\omega_2 + \frac{\partial \gamma}{\partial \theta_3}\omega_3, \tag{3.79}$$

where

$$\omega_1 = \frac{d\theta_1}{dt}, \quad \omega_2\frac{d\theta_2}{dt}, \quad \omega_3 = \frac{d\theta_3}{dt},$$

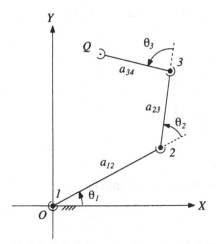

Figure 3.20 A 3R serial manipulator.

and

$$\frac{\partial x_Q}{\partial \theta_1} = -(a_{12}s_1 + a_{23}s_{1+2} + a_{34}s_{1+2+3}),$$

$$\frac{\partial x_Q}{\partial \theta_2} = -(a_{23}s_{1+2} + a_{34}s_{1+2+3}), \quad \frac{\partial x_Q}{\partial \theta_3} = -a_{34}s_{1+2+3},$$

$$\frac{\partial y_Q}{\partial \theta_1} = (a_{12}c_1 + a_{23}c_{1+2} + a_{34}c_{1+2+3}),$$

$$\frac{\partial y_Q}{\partial \theta_2} = (a_{23}c_{1+2} + a_{34}c_{1+2+3}), \quad \frac{\partial y_Q}{\partial \theta_3} = a_{34}c_{1+2+3},$$

$$\frac{\partial \gamma}{\partial \theta_1} = \frac{\partial \gamma}{\partial \theta_2} = \frac{\partial \gamma}{\partial \theta_3} = 1. \tag{3.80}$$

Equations 3.77–3.79 can be expressed in the matrix form

$$\begin{bmatrix} v_{Qx} \\ v_{Qy} \\ \omega \end{bmatrix} (= \hat{T}^*) = J^* \begin{bmatrix} \omega_1 \\ \omega_2 \\ \omega_3 \end{bmatrix} = J^*\gamma, \tag{3.81}$$

where ω is the instantaneous rotational speed of link a_{34} relative to the fixed frame; v_{Qx} and v_{Qy} are the X and Y components of the translational velocity of a point Q on a link a_{34}; ω_1, ω_2, and ω_3 are the instantaneous joint angular velocities. Also,

$$
J^* = \begin{bmatrix} \dfrac{\partial x_Q}{\partial \theta_1} & \dfrac{\partial x_Q}{\partial \theta_2} & \dfrac{\partial x_Q}{\partial \theta_3} \\[2ex] \dfrac{\partial y_Q}{\partial \theta_1} & \dfrac{\partial y_Q}{\partial \theta_2} & \dfrac{\partial y_Q}{\partial \theta_3} \\[2ex] 1 & 1 & 1 \end{bmatrix}
\tag{3.82}
$$

is known as the Jacobian matrix.

This is clearly different from the previous formulation,

$$
\hat{T} = J\gamma,
\tag{eq. 3.46}
$$

where, from (3.48), for the reference point O chosen coincident with the first joint, the Jacobian matrix is

$$
J = \begin{bmatrix} 0 & y_2 & y_3 \\ 0 & -x_2 & -x_3 \\ 1 & 1 & 1 \end{bmatrix}.
\tag{3.83}
$$

In (3.83), (x_2, y_2) and (x_3, y_3) are the coordinates of the points 2 and 3, and the columns of J are the coordinates of the joint rotor axes. A rotor with coordinates

$$
\hat{T} = \begin{bmatrix} v_{0x} \\ v_{0y} \\ \omega \end{bmatrix}
$$

quantifies the instant motion of the end effector, where (v_{0x}, v_{0y}) are the x and y components of the velocity of a point in the end effector coincident with a reference point O, which, for this case, is coincident with the first joint axis. In (3.81)

$$
\hat{T}^* = \begin{bmatrix} v_{Qx} \\ v_{Qy} \\ \omega \end{bmatrix}
$$

quantifies the instant motion of the end effector using the components of the velocity of a point Q. However, in this formulation $\hat{T}^*$ does not represent the coordinates of a rotor, and the columns of J^* do not represent the joint rotor axes coordinates.

The two results are easy to compare by expressing the partial derivatives in (3.80) in terms of the coordinates of points 2, 3, and Q, and from (3.74), (3.75), and

$$\frac{\partial x_Q}{\partial \theta_1} = -y_Q, \quad \frac{\partial x_Q}{\partial \theta_2} = -(y_Q - y_2), \quad \frac{\partial x_Q}{\partial \theta_3} = -(y_Q - y_3),$$

$$\frac{\partial y_Q}{\partial \theta_1} = x_Q, \quad \frac{\partial y_Q}{\partial \theta_2} = x_Q - x_2, \quad \frac{\partial y_Q}{\partial \theta_3} = x_Q - x_3. \tag{3.84}$$

The substitution of (3.84) into (3.77) and (3.78) yields

$$v_{Qx} = -y_Q\omega_1 - (y_Q - y_2)\omega_2 - (y_Q - y_3)\omega_3, \tag{3.85}$$

$$v_{Qy} = x_Q\omega_1 + (x_Q - x_2)\omega_2 + (x_Q - x_3)\omega_3. \tag{3.86}$$

Regroup the terms on the right sides of (3.85) and (3.86), substitute $\omega_1 + \omega_2 + \omega_3 = \omega$, and re-arrange to yield

$$v_{Qx} + \omega y_Q = y_2\omega_2 + y_3\omega_3, \tag{3.87}$$

$$v_{Qy} - \omega x_Q = -x_2\omega_2 - x_3\omega_3. \tag{3.88}$$

The instantaneous center of rotation of the end effector is denoted by point G (see Fig. 3.21). Therefore,

$$\mathbf{v}_Q = \boldsymbol{\omega} \times \mathbf{r}_Q, \quad \mathbf{v}_o = \boldsymbol{\omega} \times \mathbf{r}, \tag{3.89}$$

and

$$\mathbf{v}_Q - \mathbf{v}_o = \boldsymbol{\omega} \times (\mathbf{r}_Q - \mathbf{r}) = \boldsymbol{\omega} \times \mathbf{R}_Q, \tag{3.90}$$

or

$$\mathbf{v}_o = \mathbf{v}_Q - \boldsymbol{\omega} \times \mathbf{R}_Q = \mathbf{v}_Q - \omega \mathbf{k} \times (x_Q\mathbf{i} + y_Q\mathbf{i})$$

$$= \mathbf{v}_Q + \omega(y_Q\mathbf{i} - x_Q\mathbf{i}). \tag{3.91}$$

The scalar components of this equation are

$$v_{ox} = v_{Qx} + \omega y_Q, \tag{3.92}$$

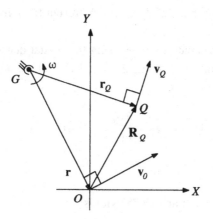

Figure 3.21 Motion of the end effector about instantaneous center G.

and

$$v_{oy} = v_{Qy} - \omega x_Q. \tag{3.93}$$

The substitution of these results into the left side of (3.87) and (3.88) yields

$$v_{ox} = y_2\omega_2 + y_3\omega_3, \tag{3.94}$$

$$v_{oy} = -x_2\omega_2 - x_3\omega_3, \tag{3.95}$$

which essentially reduces the coordinates $\hat{T}^*$ to $\hat{T}$ and the Jacobian J^* to J.

4

Dualities of parallel and serial devices

4.1   Introduction

The statics of a parallel manipulator with three parallel *RPR* kinematic chains and the first-order kinematics of a serial manipulator with three revolute joints were studied in Chapters 2 and 3. The force equation for the three connectors of a parallel manipulator and the equation for the instantaneous motion of the end effector of a serial manipulator are, respectively,

$$\hat{w} = j\lambda, \qquad \text{(eq. 2.67)}$$

and

$$\hat{T} = J\gamma. \qquad \text{(eq. 3.46)}$$

The vectors λ and γ are 3×1 column matrices of the magnitudes of the connector forces

$$\begin{bmatrix} f_1 \\ f_2 \\ f_3 \end{bmatrix}$$

and the joint angular speeds

$$\begin{bmatrix} \omega_1 \\ \omega_2 \\ \omega_3 \end{bmatrix},$$

and j, J are, respectively, 3×3 matrices, whose columns are the line coordinates of the connectors and joint axes, respectively.

117

Correspondingly, the axis coordinates for an infinitesimal rotation of the end effector of a serial manipulator can be expressed in the form

$$\delta \hat{D} = J \, \delta \theta, \qquad \text{(eq. 3.50)}$$

where the vector $\delta \theta$ is a 3×1 column matrix of the infinitesimal joint displacements,

$$\delta \theta = \begin{bmatrix} \delta \theta_1 \\ \delta \theta_2 \\ \delta \theta_3 \end{bmatrix}.$$

Dividing the left and right sides of this equation by a small time increment δt and writing

$$\hat{T} = \lim_{\delta t \to 0} \frac{\delta \hat{D}}{\delta t} \quad \text{and} \quad \omega_i = \lim_{\delta t \to 0} \frac{\delta \theta_i}{\delta t} \quad (i = 1, 2, 3)$$

yields (3.46).

The analogy, or rather the duality, between statics and instantaneous kinematics (see Chapter 3) was stated as follows: In statics a directed line segment represents the "rectilinear" concept of a force, whereas in kinematics a directed line segment represents the "circular" concept of a rotor. Clearly, (2.60) and (3.40) can be considered to be dual:

$$\hat{w} = f_1 \hat{s}_1 + f_2 \hat{s}_2 + f_3 \hat{s}_3$$

$$\hat{T} = \omega_1 \hat{S}_1 + \omega_2 \hat{S}_2 + \omega_3 \hat{S}_3.$$

It remains to perform a static analysis for a serial manipulator and dually to determine the instantaneous motion of a parallel manipulator. Before proceeding with this, it is necessary to explain what is meant by the mutual moment of a pair of lines and how this quantity is intimately related to the statics and kinematics of a rigid lamina.

4.2   Mutual moment, instantaneous power, and instantaneous work

Figure 4.1 illustrates a line $\$_i$ in the xy plane with ray coordinates

$$\hat{s} = \begin{bmatrix} c_i \\ s_i \\ r_i + r_i' \end{bmatrix}. \qquad (4.1)$$

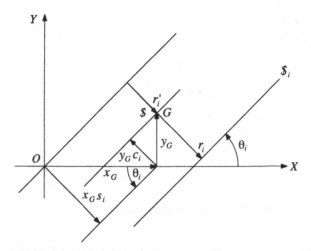

Figure 4.1 Mutual moment of two lines.

Assume that there is a second line $, drawn through the point G perpendicular to the xy plane with axis coordinates

$$\hat{S} = \begin{bmatrix} y_G \\ -x_G \\ 1 \end{bmatrix}. \tag{4.2}$$

The *mutual moment* is defined as $\hat{s}_i^T \hat{S}$ ($= \hat{S}^T \hat{s}_i$). From (4.1) and (4.2),

$$\hat{s}_i^T \hat{S} = [c_i, \, s_i; \, r_i + r_i'] \begin{bmatrix} y_G \\ -x_G \\ 1 \end{bmatrix}$$

$$= r_i + r_i' - (x_G s_i - y_G c_i). \tag{4.3}$$

From Figure 4.1, $(x_G s_i - y_G c_i) = r_i'$ and therefore

$$\hat{s}_i^T \hat{S} = r_i. \tag{4.4}$$

The mutual moment for this pair of normalized lines, which are mutually perpendicular, is their common perpendicular distance r_i. The mutual moment can be considered to be an invariant in that the common perpendicular distance between a given pair of lines is the same no matter how a coordinate system is chosen. The product $\hat{s}_i^T \hat{S}$ ($= r_i$) is invariant under the group of Eu-

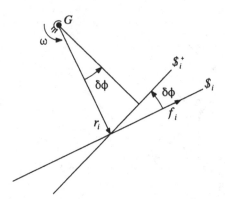

Figure 4.2 Instantaneous power generated by a force.

clidean motions. Also, if one applies a force of magnitude f_i on either line, then the moment about the other is $f_i r_i$.

Assume now a force of magnitude f_i on the line $\$_i$ which acts upon a lamina that is modeled in Figure 4.2 by a rigid link r_i connected to the ground via a revolute joint at G, the link rotating with an instantaneous angular velocity $\omega = \lim\limits_{\delta t \to 0} \dfrac{\delta \phi}{\delta t}$. The instantaneous power generated by the force is given by

$$\hat{w}^T \hat{T} = (f_i \hat{s}_i^T)\,(\omega \hat{S}) = f_i \omega \hat{s}_i^T \hat{S} = f_i \omega r_i, \tag{4.5}$$

and, analogously, the instantaneous work is given by

$$\hat{w}^T \delta \hat{D} = (f_i \hat{s}_i^T)(\delta \phi \hat{S}) = f_i \delta \phi \hat{s}_i^T \hat{S} = f_i \delta \phi r_i. \tag{4.6}$$

These simple results demonstrate that not only are the quantities instantaneous power and instantaneous work invariant with respect to the Euclidean group of motions, which they must be, but they are also intimately related to the geometry of lines.

It is important to note that when $r_i = 0$, the mutual moment is zero and the instantaneous power and instantaneous work are also zero. This means that if the line of action of the force intersects the axis of rotation then the force cannot produce motion no matter how great the intensity of the force, f_i. *Then the pair of lines is reciprocal to one another just as the force and rotor are reciprocal.*

Figure 4.3 illustrates a lamina which has a single twist of freedom about

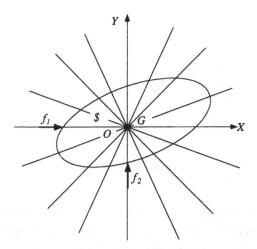

Figure 4.3 A rotor and a reciprocal two-system of forces.

a rotor axis through G perpendicular to the xy plane, the coordinates of which are

$$\hat{T} = \omega \begin{bmatrix} 0 \\ 0 \\ 1 \end{bmatrix}.$$

The system of forces reciprocal to this one-system rotor is a two-system. It is possible to select any pair of forces in the pencil, for example, forces with magnitudes f_1 and f_2, which act on lines through G with $\theta_1 = 0$ and $\theta_2 = \pi/2$. Any force in the pencil with coordinates

$$\hat{w} = f_1 \begin{bmatrix} 1 \\ 0 \\ 0 \end{bmatrix} + f_2 \begin{bmatrix} 0 \\ 1 \\ 0 \end{bmatrix}$$

is clearly reciprocal to the rotor because

$$\hat{w}^T \hat{T} = [f_1, f_2; 0] \begin{bmatrix} 0 \\ 0 \\ \omega \end{bmatrix} = 0.$$

From a physical standpoint any force that acts upon the lamina and passes through G cannot produce motion.

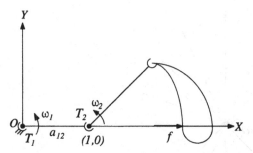

Figure 4.4 Two rotors and a reciprocal one-system.

Consider now a two-system of twists T_1 and T_2, a serial 2R manipulator (see Fig. 4.4). Without loss of generality, the axes of the twists $\$_1$ and $\$_2$ are located on the X axis. Their coordinates are, respectively,

$$\omega_1 \begin{bmatrix} 0 \\ 0 \\ 1 \end{bmatrix} \quad \text{and} \quad \omega_2 \begin{bmatrix} 0 \\ -a_{12} \\ 1 \end{bmatrix}.$$

The lamina has two degrees of freedom and the instant motion of the end effector is any linear combination,

$$\hat{T} = \omega_1 \begin{bmatrix} 0 \\ 0 \\ 1 \end{bmatrix} + \omega_2 \begin{bmatrix} 0 \\ -a_{12} \\ 1 \end{bmatrix}.$$

Any resulting axis of rotation must pass through the X axis because

$$\hat{T} = \begin{bmatrix} 0 \\ -\omega_2 a_{12} \\ (\omega_1 + \omega_2) \end{bmatrix} = (\omega_1 + \omega_2) \begin{bmatrix} 0 \\ \dfrac{-\omega_2 a_{12}}{(\omega_1 + \omega_2)} \\ 1 \end{bmatrix},$$

i.e., $x = \omega_2 a_{12} / (\omega_1 + \omega_2)$, $y = 0$. Because there are two independent freedoms in the plane there must be a single force of constraint which cannot produce motion of the end effector. This is a force of any magnitude f whose line of action passes through the axes of the two rotors (see Fig. 4.4). From a physical standpoint, one can imagine that the end effector extends over the

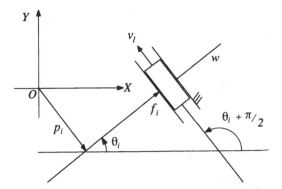

Figure 4.5 Instantaneous sliding motion.

X axis. The force w cannot produce motion of the end effector regardless of the magnitude of f. The coordinates of the constraint force w are

$$\hat{w} = f \begin{bmatrix} 1 \\ 0 \\ 0 \end{bmatrix}$$

and $\hat{w}^{\mathrm{T}} \hat{T} = 0$.

Assume now that a body instantaneously has a pure sliding motion. This can be modeled by a prismatic joint (see Fig. 4.5). The slider has a single freedom. The system of forces reciprocal to this one-system of freedom is a two-system which consists of any force w with coordinates

$$\hat{w} = f_i \begin{bmatrix} c_i \\ s_i \\ p_i \end{bmatrix}$$

in the parallel pencil of forces perpendicular to the sliding motion T, which has coordinates

$$\hat{T} = v_i \begin{bmatrix} -s_i \\ c_i \\ 0 \end{bmatrix},$$

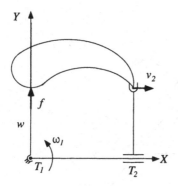

Figure 4.6 An *RP* manipulator and its reciprocal one-system.

together with a pure couple with coordinates

$$c_0 \begin{bmatrix} 0 \\ 0 \\ 1 \end{bmatrix}.$$

Any linear combination

$$\hat{w} = f_i \begin{bmatrix} c_i \\ s_i \\ p_i \end{bmatrix} + c_0 \begin{bmatrix} 0 \\ 0 \\ 1 \end{bmatrix}$$

cannot produce motion, and w and T are reciprocal because $\hat{w}^T \hat{T} = 0$.

Consider a two-system of twists, T_1 a rotor, and T_2 a pure translation. This can be modeled without any loss of generality by an RP manipulator (see Fig. 4.6). The coordinates of T_1 and T_2 are, respectively,

$$\hat{T}_1 = \omega_1 \begin{bmatrix} 0 \\ 0 \\ 1 \end{bmatrix} \quad \text{and} \quad \hat{T}_2 = v_2 \begin{bmatrix} 1 \\ 0 \\ 0 \end{bmatrix}.$$

The lamina has two independent freedoms and the instant motion of the end-effector is a linear combination,

$$\hat{T} = \omega_1 \begin{bmatrix} 0 \\ 0 \\ 1 \end{bmatrix} + v_2 \begin{bmatrix} 1 \\ 0 \\ 0 \end{bmatrix}.$$

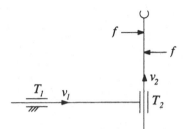

Figure 4.7 Displacements T_1, T_2, and a reciprocal one-system.

There must be a single force, a one-system reciprocal to this two-system. This must be the force in the planar pencil of parallel forces reciprocal to T_2 which passes through the rotor axis of T_1, the coordinates for which are

$$\hat{w} = f \begin{bmatrix} 0 \\ 1 \\ 0 \end{bmatrix}.$$

The reciprocity condition $\hat{w}^T \hat{T} = 0$ is satisfied. From a physical standpoint, one can imagine that the end effector extends over the y axis. So the force w cannot produce motion of the end effector regardless of the magnitude of f.

Finally, consider a two-system that consists of pure translations T_1 and T_2. This can be modeled by a PP manipulator (see Fig. 4.7). The coordinates for T_1 and T_2 are, respectively,

$$\hat{T}_1 = v_1 \begin{bmatrix} 1 \\ 0 \\ 0 \end{bmatrix} \quad \text{and} \quad \hat{T}_2 = v_2 \begin{bmatrix} 0 \\ 1 \\ 0 \end{bmatrix}.$$

The instant motion of the end effector is a pure translation and

$$\hat{T} = v_1 \begin{bmatrix} 1 \\ 0 \\ 0 \end{bmatrix} + v_2 \begin{bmatrix} 0 \\ 1 \\ 0 \end{bmatrix}.$$

There is a one-system, a pure couple with coordinates

$$\hat{w} = c \begin{bmatrix} 0 \\ 0 \\ 1 \end{bmatrix},$$

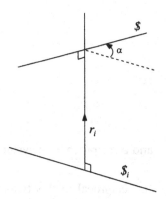

Figure 4.8 A pair of skew lines.

which is reciprocal to this nonrotating assemblage and $\hat{w}^T \, \hat{T} = 0$. From a physical standpoint, it is clear that a pair of equal and opposite forces (see Fig. 4.7) applied to the end effector can produce no motion.

In summary, for the motion of a lamina in a plane, the dimension of the space of twists of freedom (dim T) and the dimension of the space of reciprocal forces (dim w), which are commonly defined as forces of constraint, add up to three, dim T + dim w = 3. When the lamina has three independent freedoms dim T = 3 and thus dim w = 0, i.e., there are no constraints acting upon the lamina. For a lamina with two independent freedoms, dim T = 2, dim w = 1, and thus there is a single constraint, a single force reciprocal to the two-system. For a lamina with one freedom, dim T = 1, dim w = 2, and there is a two-system of constraints or a two-system of forces reciprocal to the one freedom. When there is no motion, the lamina is completely constrained, dim w = 3 and dim T = 0.

Finally, it is interesting to note that the mutual moment of a pair of skew lines \$ and $\$_i$ is given by

$$\hat{s}_i^T \hat{S} = \hat{S}^T \hat{s}_i = -r_i \sin \alpha. \tag{4.7}$$

Here (see Fig. 4.8) r_i and α are, respectively, the common perpendicular and the twist angle between the lines. If the Plücker coordinates for the lines $\$_i$ and \$ are $\hat{s}_i = [L_i, M_i, N_i; P_i, Q_i, R_i]^T$ and $\hat{S} = [P, Q, R; L, M, N]^T$, then

$$\hat{s}_i^T \hat{S} = \hat{S}^T \, \hat{s}_i = L_i P + M_i Q + N_i R + L P_i + M Q_i + N R_i. \tag{4.8}$$

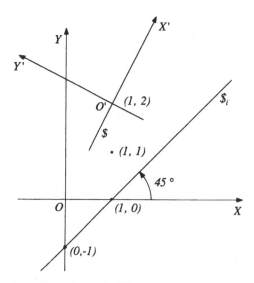

Exercise Figure 4.1(1)

—— **EXERCISE 4.1** ————————————

1. (a) Compute the mutual moment of the line $\$_i$ in the *xy* plane and the line $\$$ perpendicular to the *xy* plane through the point (1, 1) (see Exr. Fig. 4.1(1)).

(b) Compute the mutual moment of the two lines when *O* is located at the points (1, 0) and (0, 1).

(c) The origin of the coordinate system is translated to the point (1, 2) and the coordinate system is rotated 60 degrees anticlockwise. Determine the Plücker coordinates of the two lines in the new coordinate system and compute the mutual moment.

2. (a) The wheel that is rotating anticlockwise (see Exr. Fig. 4.1(2)) at an angular speed of 10 rads/sec. experiences an impulsive force of 5 lbf. Compute the instant power gained by the system. Repeat the calculation for a clockwise rotation of 10 rads/sec. and comment on your results.

(b) The two-system of constraints can be quantified by the base forces with magnitudes f_1 and f_2 acting along the *X* and *Y* axes as shown in the figure. Show that the impulsive force cannot be expressed as a linear combination of the base forces.

3. Compute the instant motion of the end effector of the 2*R* manipulator (i.e., locate the instant center and compute the angular speed) when $\omega_1 = 2$ rads/sec.,

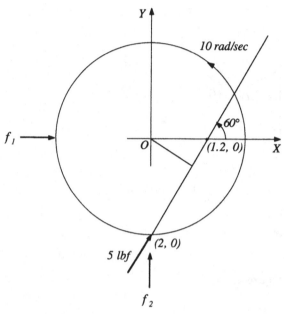

Exercise Figure 4.1(2)

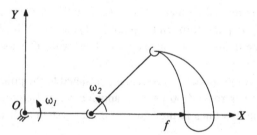

Exercise Figure 4.1(3)

$\omega_2 = 8$ rads/sec.; $\omega_1 = -2$ rads/sec., $\omega_2 = -8$ rads/sec.; $\omega_1 = -2$ rads/sec., $\omega_2 = 8$ rads/sec.; $\omega_1 = -8$ rads/sec. $\omega_2 = 2$ rads/sec. that is, show that the instant center must lie on the x axis and that any force along the x axis is reciprocal to the end-effector motion. Use Exercise Figure 4.1(3).

4.   **(a)** The end effector maintains a point contact at P such that it is constrained to move along the X axis, as illustrated in Exercise Figure 4.1(4a). Determine dim T and dim w, and write a set of coordinates for the bases of the twist

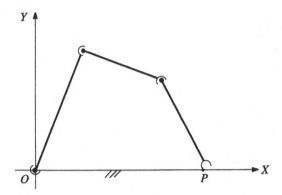

Exercise Figure 4.1(4a)

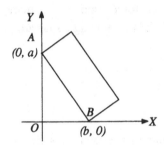

Exercise Figure 4.1(4b)

space and the corresponding reciprocal space. Show that the coordinates you have chosen satisfy reciprocity.

(b) The beam maintains contact with the Y and the X axes at points A and B, as shown in Exercise Figure 4.1(4b). Determine dim T and dim w, and write a set of coordinates for the bases of the twist space and the corresponding reciprocal space. Show that the coordinates you have chosen satisfy reciprocity.

(c) The beam maintains contact at points A, B, and D (see Exr. Fig. 4.1(4c)). Determine dim T and dim w, and give a set of coordinates for the bases of the twist space and the corresponding reciprocal space.

4.3 A static analysis of a planar serial manipulator

Consider a force with magnitude f that acts upon the end effector of a planar 3R manipulator on a line $\$$ with ray coordinates $\hat{s}$. The axis coordinates of the lines of the rotors $\$_1$, $\$_2$, and $\$_3$ are $\hat{S}_1$, $\hat{S}_2$, and $\hat{S}_3$. This force

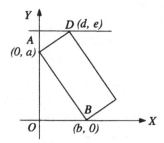

Exercise Figure 4.1(4c)

will induce a resultant torque τ_i at each joint, and at each joint the motor must generate an equilibrating torque $-\tau_i$ to maintain equilibrium (see Fig. 4.9).

The resultant torques due to the applied force f alone (i.e., disregarding any effects of either gravity or the inertia of the moving parts) are given by

$$\tau_1 = fr_1,$$

$$\tau_2 = fr_2,$$

$$\tau_3 = fr_3. \tag{4.9}$$

Now r_1, r_2 and r_3 are the mutual moments of the rotor axes and the line $\$$. Therefore

$$\tau_1 = f\,\hat{S}_1^T \hat{s} = \hat{S}_1^T \hat{w},$$

$$\tau_2 = f\,\hat{S}_2^T \hat{s} = \hat{S}_2^T \hat{w},$$

$$\tau_3 = f\,\hat{S}_3^T \hat{s} = \hat{S}_3^T \hat{w}. \tag{4.10}$$

Equation 4.10 can be expressed in matrix form as

$$\tau = J^T \hat{w}, \tag{4.11}$$

where

$$\tau = \begin{bmatrix} \tau_1 \\ \tau_2 \\ \tau_3 \end{bmatrix} \quad \text{and} \quad J = \begin{bmatrix} y_1 & y_2 & y_3 \\ -x_1 & -x_2 & -x_3 \\ 1 & 1 & 1 \end{bmatrix}. \qquad \text{(see (3.47))}$$

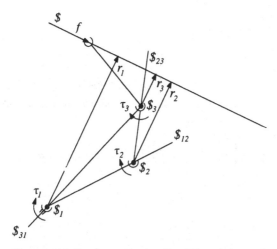

Figure 4.9 Static analysis of a 3R serial manipulator.

When a force is applied to the end effector it is a simple matter to compute the magnitudes of the resultant torque at each joint, and, subsequently, the equilibrants $-\tau_1$, $-\tau_2$, and $-\tau_3$. This is called *reverse static analysis.* Conversely, when the actual torques generated by the motors are known we need to compute the actual force applied to the end effector. This is called *forward static analysis.*

The joint torque/force relationship, equation 4.11, can also be derived using instant power. Assume that the end effector is moving instantaneously on a rotor with coordinates $\hat{T}$ and that the corresponding joint angular velocities are ω_1, ω_2, and ω_3. Equate the instant power generated at the joints $\omega_1\tau_1 + \omega_2\tau_2 + \omega_3\tau_3$, or in matrix form $\gamma^T \tau$, to the instant power generated at the end effector $\hat{T}^T \hat{w}$, which yields $\gamma^T\tau = \hat{T}^T\hat{w}$. Furthermore, from (3.46), $\hat{T}^T = \gamma^T J^T$. Substituting this result into the right side of $\gamma^T\tau = \hat{T}^T\hat{w}$ and premultiplying both sides by $(\gamma^T)^{-1}$ yields (4.11).

Further, from (4.11), the coordinates for the force for a forward static analysis are given by

$$\hat{w} = (J^T)^{-1}\tau. \tag{4.12}$$

The geometrical meaning of J^{-1} (referred to in Chapter 3) was shown to be

$$J^{-1} = \begin{bmatrix} a_{23}\hat{s}_{23}^T \\ a_{31}\hat{s}_{31}^T \\ a_{12}\hat{s}_{12}^T \end{bmatrix} / \det J, \qquad \text{(eq. 3.53)}$$

where $\hat{s}_{12}$, $\hat{s}_{23}$, and $\hat{s}_{31}$ are the coordinates of the lines $\$_{12}$, $\$_{23}$, and $\$_{31}$ joining the pivots 1–2, 2–3, and 3–1, respectively, as shown in Figure 4.9. Now

$$(J^T)^{-1} = (J^{-1})^T = \begin{bmatrix} a_{23}\hat{s}_{23}^T \\ a_{31}\hat{s}_{31}^T \\ a_{12}\hat{s}_{12}^T \end{bmatrix}^T / \det J = [a_{23}\hat{s}_{23} \ a_{31}\hat{s}_{31} \ a_{12}\hat{s}_{12}] / \det J.$$

Thus from (4.12):

$$\hat{w} = [a_{23}\hat{s}_{23} \ a_{31}\hat{s}_{31} \ a_{12}\hat{s}_{12}] \begin{bmatrix} \tau_1 \\ \tau_2 \\ \tau_3 \end{bmatrix} / \det J. \qquad (4.13)$$

We also show (in Chapter 3) that det J can be expressed in three ways:

$$\det J = a_{23}\hat{s}_{23}^T \ \hat{S}_1, \qquad \text{(eq. 3.55)}$$

$$\det J = a_{31}\hat{s}_{31}^T \ \hat{S}_2, \qquad \text{(eq. 3.56)}$$

$$\det J = a_{12}\hat{s}_{12}^T \ \hat{S}_3. \qquad \text{(eq. 3.57)}$$

The geometrical meaning of det J is now apparent. Equations 3.55, 3.56, and 3.57 are alternative expressions for twice the area of the triangle formed by the three turning joints. For example, (3.55) is the product of the side a_{23} by the perpendicular distance $\hat{s}_{23}^T \hat{S}_1$ from the first joint to side a_{23}.

Assume now that the end effector is in contact with a rigid body. Individually applied motor torques

$$\begin{bmatrix} \tau_1 \\ 0 \\ 0 \end{bmatrix}, \begin{bmatrix} 0 \\ \tau_2 \\ 0 \end{bmatrix}, \text{ and } \begin{bmatrix} 0 \\ 0 \\ \tau_3 \end{bmatrix}$$

cause the end effector to apply contact forces to the body. The coordinates of these forces can be obtained by substituting, in turn, these motor torques into the right side of (4.13). After this procedure, the introduction of the corresponding expression (3.55), (3.56), and (3.57) for det J yields

$$\hat{w} \equiv \hat{w}_{23} = \hat{s}_{23}\tau_1/\hat{s}_{23}^T\hat{S}_1 = f_{23}\hat{s}_{23} \tag{4.14}$$

$$\hat{w} \equiv \hat{w}_{31} = \hat{s}_{31}\tau_2/\hat{s}_{31}^T\hat{S}_2 = f_{31}\hat{s}_{31}, \tag{4.15}$$

$$\hat{w} \equiv \hat{w}_{12} = \hat{s}_{12}\tau_3/\hat{s}_{12}^T\hat{S}_3 = f_{12}\hat{s}_{12}, \tag{4.16}$$

where

$$\frac{\tau_1}{\hat{s}_{23}^T\hat{S}_1} = f_{23}, \tag{4.17}$$

$$\frac{\tau_2}{\hat{s}_{31}^T\hat{S}_2} = f_{31}, \tag{4.18}$$

$$\frac{\tau_3}{\hat{s}_{12}^T\hat{S}_3} = f_{12}. \tag{4.19}$$

Hence, individually applied motor torques τ_1, τ_2, and τ_3 cause the end effector to apply forces to the body along the lines $\$_{23}$, $\$_{31}$, and $\$_{12}$, respectively.

Consider again the reverse kinematic solution for the twist equation for the 3R manipulator,

$$\hat{T} = \omega_1 \hat{S}_1 + \omega_2 \hat{S}_2 + \omega_3 \hat{S}_3. \tag{eq. 3.40}$$

Solutions for ω_1, ω_2, and ω_3 can be obtained directly from (3.40) by forming the reciprocal products of the left and right sides with the coordinates $\hat{s}_{23}$, $\hat{s}_{31}$, and $\hat{s}_{12}$ of the lines $\$_{23}$, $\$_{31}$, and $\$_{12}$, which yields, respectively,

$$\omega_1 = \frac{\hat{s}_{23}^T\hat{T}}{\hat{s}_{23}^T\hat{S}_1}, \tag{eq. 3.59}$$

$$\omega_2 = \frac{\hat{s}_{31}^T\hat{T}}{\hat{s}_{31}^T\hat{S}_2}, \tag{eq. 3.60}$$

$$\omega_3 = \frac{\hat{s}_{12}^T\hat{T}}{\hat{s}_{12}^T\hat{S}_3}. \tag{eq. 3.61}$$

In forming the reciprocal product of (3.40) with $\hat{s}_{23}$, the products $\hat{s}_{23}^T \hat{S}_2$ and $\hat{s}_{23}^T \hat{S}_3$ vanish. The line $\$_{23}$ intersects both the lines $\$_2$ and $\$_3$ (see Fig. 4.9). Analogously, $\hat{s}_{31}^T \hat{S}_1$ and $\hat{s}_{31}^T \hat{S}_3$ vanish, and $\hat{s}_{12}^T \hat{S}_1$ and $\hat{s}_{12}^T \hat{S}_2$ vanish, because the line $\$_{31}$ instersects both the lines $\$_1$ and $\$_2$, and the line $\$_{12}$ intersects both the lines $\$_1$ and $\$_2$.

──────── **EXERCISE 4.2** ────────────────

1. Exercise Figure 4.2(1) illustrates two positions of a 3*R* manipulator. All the linear dimensions are in feet.

(a) A force of 5 lbf acts on the end effector as shown in the figure. Compute the corresponding resultant torques for the two positions and the equilibrating joint torques.

(b) Assume that the end effector is in contact with some object and is in equilibrium with equilibrating joint torques $\tau_1 = 2$, $\tau_2 = 3$, and $\tau_3 = 4$ (lbf = ft). Compute the corresponding external forces that act upon the end effector for the two positions.

2. The end effector of the 3*R* manipulator is instantaneously rotating about point *C* (0, 2), as illustrated in Exercise Figure 4.2(2). The velocity of point *P* is $v_p = 1$ in./sec. Determine the angular velocity of the lamina and the velocity v_o of a point in the lamina coincident with the reference point *O*. Compute the corresponding values for the joint speeds ω_1, ω_2, and ω_3 from the twist equation

$$\hat{T} = \omega_1 \hat{S}_1 + \omega_2 \hat{S}_2 + \omega_3 \hat{S}_3$$

by forming reciprocal products with the coordinates of lines $\$_{23}$, $\$_{31}$, and $\$_{12}$, respectively. Compare your results with part 1 of Exercise 3.2.

3. We require that the end effector of the *RPR* manipulator rotate on an instant center with coordinates (2, 1) with a clockwise angular speed of 10 rads/sec. Compute ω_1, v_2, and ω_3 by forming in turn reciprocal products of the twist equation with three lines; each line is reciprocal to pairs of joint motions, as shown in Exercise Figure 4.2(3).

4. We show for a 3*R* manipulator that det $J = a_{12}\hat{s}_{12}^T \hat{S}_3$ (see (3.57)), which is twice the area of the triangle 123 shown in Exercise Figure 4.2(4).

(a) Deduce that det $J = a_{12} a_{23} \sin \theta_2$.

(b) Deduce that the maximum value for det *J* is det $J_{max} = a_{12}a_{23}$ and $\lambda = \det J / \det J_{max} = \sin \theta_2$.

(c) Sketch the graph $\lambda = \sin \theta_2$ for $0 \leq \theta_2 \leq 360$ degrees. Draw the pair configurations of the manipulator for $\lambda = 0$ and $\lambda = \pm 1$. (Note that the ratio λ is independent of the manipulator's dimensions.)

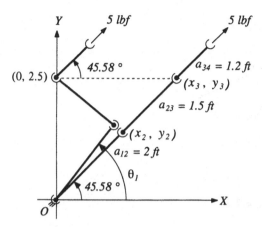

Exercise Figure 4.2(1)

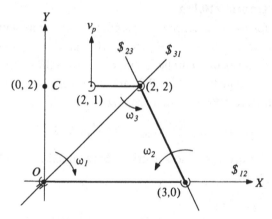

Exercise Figure 4.2(2)

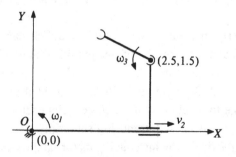

Exercise Figure 4.2(3)

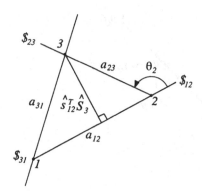

Exercise Figure 4.2(4)

4.4   A static analysis of a parallel manipulator with *RPR* and 3*R* kinematic chains

The skeletal forms of planar motion parallel manipulators with *RPR* and 3*R* serial chains are illustrated in Figures 4.10 and 4.11. Before proceeding with a static analysis of a parallel manipulator with 3*R* serial chains it is interesting to restate equations 2.88–2.90 for the reverse static analysis of a manipulator with three *RPR* serial chains, viz.,

$$f_1 = \hat{S}_{23}^T \hat{w} / \hat{S}_{23}^T \hat{s}_1, \quad f_2 = \hat{S}_{31}^T \hat{w} / \hat{S}_{31}^T \hat{s}_2, \quad f_3 = \hat{S}_{12}^T \hat{w} / \hat{S}_{12}^T \hat{s}_3,$$

which were obtained by inverting the *j* matrix. These equations can be obtained directly from (2.60),

$$\hat{w} = f_1 \hat{s}_1 + f_2 \hat{s}_2 + f_3 \hat{s}_3,$$

by forming, in turn, the reciprocal products of the left and right sides of (2.60) with the coordinates $\hat{S}_{23}$, $\hat{S}_{31}$, and $\hat{S}_{12}$ of the lines $\$_{23}$, $\$_{31}$, and $\$_{12}$ (see Figure 4.10).

It is now possible to perform a static analysis of a parallel manipulator with three 3*R* serial chains using the results of Section 4.3. Admittedly, the notation can become tedious because it is desirable to label the sequence of joints in a serial chain by 1, 2, 3, beginning with the grounded joint. Furthermore, it is also desirable to label each of the three serial chains by 1, 2, 3. This is accomplished in Fig. 4.11 by introducing subscripts.

Assume that we actuate the device by introducing motors at the joints $(1)_1$,

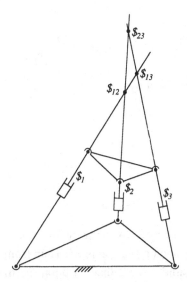

Figure 4.10 Parallel device with *RPR* serial chains.

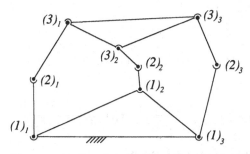

Figure 4.11 Parallel device with *3R* serial chains.

$(1)_2$, and $(1)_3$ and that torques $(\tau_1)_1$, $(\tau_1)_2$, and $(\tau_1)_3$ are applied. Then from (4.14) and (4.17) the coordinates of the resultant force acting upon the movable lamina are given by

$$\hat{w} = (\hat{w}_{23})_1 + (\hat{w}_{23})_2 + (\hat{w}_{23})_3, \tag{4.20}$$

or

$$\hat{w} = (f_{23})_1 \, (\hat{s}_{23})_1 + (f_{23})_2 \, (\hat{s}_{23})_2 + (f_{23})_3 \, (\hat{s}_{23})_3, \tag{4.21}$$

where

$$(f_{23})_1 = \frac{(\tau_1)_1}{(\hat{s}_{23}^T)_1\,(\hat{S}_1)_1}, \quad (f_{23})_2 = \frac{(\tau_1)_2}{(\hat{s}_{23}^T)_2\,(\hat{S}_1)_2}, \quad (f_{23})_3 = \frac{(\tau_1)_3}{(\hat{s}_{23}^T)_3\,(\hat{S}_1)_3}, \quad (4.22)$$

and (4.21) can be expressed in matrix form by

$$\hat{w} = [(\hat{s}_{23})_1,\, (\hat{s}_{23})_2,\, (\hat{s}_{23})_3]\, \mathbf{f}_{23}, \tag{4.23}$$

where

$$f_{23} = \begin{bmatrix} (f_{23})_1 \\ (f_{23})_2 \\ (f_{23})_3 \end{bmatrix}.$$

Let us actuate the parallel manipulator by introducing motors at joints $(2)_1$, $(2)_2$, and $(2)_3$, and torques $(\tau_2)_1$, $(\tau_2)_2$, and $(\tau_2)_3$ are applied. Then from (4.18) and (4.15) the coordinates of the resultant force that act upon the movable lamina are given by

$$\hat{w} = (\hat{w}_{31})_1 + (\hat{w}_{31})_2 + (\hat{w}_{31})_3, \tag{4.24}$$

or

$$\hat{w} = (f_{31})_1(\hat{s}_{31})_1 + (f_{31})_2(\hat{s}_{31})_2 + (f_{31})_3(\hat{s}_{31})_3, \tag{4.25}$$

where

$$(f_{31})_1 = \frac{(\tau_2)_1}{(\hat{s}_{31}^T)_1(\hat{S}_2)_1}, \quad (f_{31})_2 = \frac{(\tau_2)_2}{(\hat{s}_{31}^T)_2(\hat{S}_2)_2}, \quad (f_{31})_3 = \frac{(\tau_2)_3}{(\hat{s}_{31}^T)_3(\hat{S}_2)_3}, \quad (4.26)$$

and (4.25) can be expressed in matrix form as

$$\hat{w} = [(\hat{s}_{31})_1,\, (\hat{s}_{31})_2,\, (\hat{s}_{31})_3]\, \mathbf{f}_{31}, \tag{4.27}$$

where

$$\mathbf{f}_{31} = \begin{bmatrix} (f_{31})_1 \\ (f_{31})_2 \\ (f_{31})_3 \end{bmatrix}.$$

Analogously, if joints $(3)_1$, $(3)_2$, and $(3)_3$ are actuated, the resultant force that acts upon the movable lamina is

$$\hat{w} = [(\hat{s}_{12})_1,\, (\hat{s}_{12})_2,\, (\hat{s}_{12})_3]\, \mathbf{f}_{12}, \tag{4.28}$$

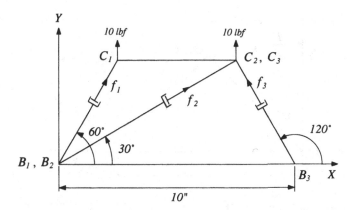

Exercise Figure 4.3(1a)

where

$$\mathbf{f}_{12} = \begin{bmatrix} (f_{12})_1 \\ (f_{12})_2 \\ (f_{12})_3 \end{bmatrix}.$$

――――― **EXERCISE 4.3** ―――――――――――――――――

1.   **(a)** Determine the unitized coordinates of the lines $\$_1$, $\$_2$, and $\$_3$ that join the connectors B_1C_1, B_2C_2, and B_3C_3. Use point $B_1(B_2)$ as your reference (see also part 1 of Exercise 2.3). Then, determine the coordinates of the points of intersection of the pairs of lines ($\$_2$, $\$_3$), ($\$_3$, $\$_1$), and ($\$_1$, $\$_2$). Hence determine the coordinates of the lines $\$_{23}$, $\$_{31}$, and $\$_{12}$ which are, respectively, reciprocal to the line pairs ($\$_2$, $\$_3$), ($\$_3$, $\$_1$), and ($\$_1$, $\$_2$). Following this, obtain expressions for the resultants f_1, f_2, and f_3 from the force equation

$$\hat{w} = f_1\hat{s}_1 + f_2\hat{s}_2 + f_3\hat{s}_3.$$

Use these expressions to compute the resultants for a vertical force of 10 lbf acting through point C_1 (see Exr. Fig. 4.3(1a)). Following this, compute the resultants for a vertical force of 10 lbf acting through point C_2 (C_3), as shown.

     (a) Repeat the previous exercise for the force of 5 lbf that acts upon the truss (see Exr. Fig. 4.3(1b)). Following this, compute the resultants when the force of 5 lbf acts vertically through point C_1 and then through point C_2 (C_3).

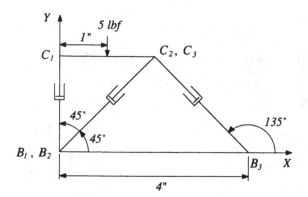

Exercise Figure 4.3(1b)

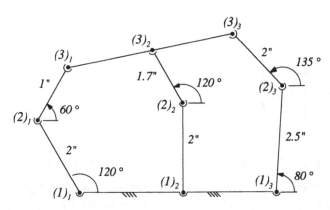

Exercise Figure 4.3(2)

2. The upper platform of the drawing in Exercise Figure 4.3(2) is in contact equilibrium with another body. Anticlockwise unit joint torques of 1 lbf inch are applied at the three base joints, the three intermediate joints, and the three upper platform joints. Compute the corresponding external forces acting upon the upper platform for each of these three cases.

4.5 A kinematic analysis of a parallel manipulator

Figure 4.12 illustrates a movable lamina connected to a fixed base via three in-parallel *RPR* kinematic chains. Assume that the prismatic pair in each chain is actuated, and that the moving platform undergoes an instantaneous rotation about some point *G*, which is called the instantaneous center

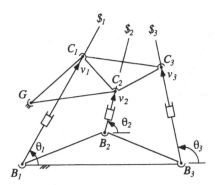

Figure 4.12 A parallel manipulator.

for the rotation. This can be modeled by a revolute joint at G (connected to the ground) to which the movable lamina is rigidly connected.

The coordinates for the instantaneous motion

$$\hat{T} = \begin{bmatrix} v_{ox} \\ v_{oy} \\ \omega \end{bmatrix}$$

are known and measured relative to some reference point 0 in the fixed base. We need to compute the sliding velocities v_1, v_2, and v_3 of the actuated prismatic pairs. Assume that the geometry of the system is known and the coordinates of the lines $\$_1$, $\$_2$, and $\$_3$ are, respectively, $\hat{s}_1$, $\hat{s}_2$, and $\hat{s}_3$.

Consider the instantaneous kinematics of any one of the *RPR* chains (see Fig. 4.13), which is labeled with the subscript i, that connects the platform to the base triangle. The instantaneous first-order kinematics of the platform is given by

$$\hat{T} = \hat{T}_{1i} + \hat{T}_{2i} + \hat{T}_{3i}, \tag{4.29}$$

which can be expressed in the form

$$\hat{T} = \omega_{1i} \hat{S}_{1i} + \hat{T}_{2i} + \omega_{3i} \hat{S}_{3i}, \tag{4.30}$$

where ω_{1i} and ω_{3i} are the angular joint speeds and $\hat{S}_{1i}$ and $\hat{S}_{3i}$ are the axis coordinates for the revolute joints, and $\hat{T}_{2i} = [v_x, v_y; 0]^T = v_i[c_i, s_i; 0]^T$ are the coordinates for the slider motion.

Compute the slider velocity v_i, which is the velocity component of the moving pivot C_i parallel to $\$_i$. This is easy to accomplish with the reciproc-

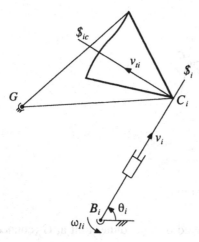

Figure 4.13 The instantaneous first-order kinematics of an *RPR* chain.

ity condition developed in Section 4.2. Clearly, line $\$_i$ intersects the axes of both revolute joints. Therefore

$$\hat{s}_i^T \hat{S}_{1i} = \hat{s}_i^T \hat{S}_{3i} = 0. \tag{4.31}$$

Now $\hat{s}_i = [c_i, s_i; p_i]^T$, and hence

$$\hat{s}_i^T \hat{T}_{2i} = v_i \, [c_i, s_i; p_i] \begin{bmatrix} c_i \\ s_i \\ 0 \end{bmatrix}$$

$$= v_i(c_i^2 + s_i^2) = v_i. \tag{4.32}$$

The formation of the reciprocal product of the left and right sides of (4.30) with $\hat{s}_i$ and the substitution of (4.31) and (4.32) yields

$$v_i = \hat{s}_i^T \hat{T}. \tag{4.33}$$

Hence, for the three connecting *RPR* chains,

$$v_1 = \hat{s}_1^T \hat{T},$$
$$v_2 = \hat{s}_2^T \hat{T},$$
$$v_3 = \hat{s}_3^T \hat{T}, \tag{4.34}$$

which can be expressed in the matrix form

$$\mathbf{v} = j^T \, \hat{T}, \tag{4.35}$$

where

$$\mathbf{v} = \begin{bmatrix} v_1 \\ v_2 \\ v_3 \end{bmatrix} \text{ and } j = \begin{bmatrix} c_1 & c_2 & c_3 \\ s_1 & s_2 & s_3 \\ p_1 & p_2 & p_3 \end{bmatrix}. \qquad \text{(see (2.68))}$$

When the instantaneous twist of the platform is specified, it is a simple matter to compute the values for the actuator speeds v_1, v_2, and v_3. This is called the *reverse or inverse velocity analysis.* Conversely, when the actuator speeds are known, then we need to compute the actual instantaneous twist of the platform. This is called the *forward velocity analysis.* From (4.35):

$$\hat{T} = (j^T)^{-1}\mathbf{v}. \tag{4.36}$$

Equation 4.36 can also be derived using instant power. Assume that a force with coordinates $\hat{w}$ is acting instantaneously on the moving platform and the corresponding forces generated in the connectors are $f_1, f_2,$ and f_3. Equate the instant power generated in the joints $\boldsymbol{\lambda}^T\mathbf{v}$ ($\boldsymbol{\lambda}^T = [f_1, f_2, f_3]$) to the instant power generated by the platform $\hat{w}^T \, \hat{T}$ to yield $\boldsymbol{\lambda}^T\mathbf{v} = \hat{w}^T \, \hat{T}$. Substituting $\hat{w}^T = \boldsymbol{\lambda}^T j^T$ (see (2.67)) and premultiplying both sides by $(\boldsymbol{\lambda}^T)^{-1}$ yields (4.35).

Chapter 2 shows that the geometric meaning of j^{-1} is

$$j^{-1} = [s_{3-2}\hat{S}_{23}^T \quad s_{1-3}\hat{S}_{31}^T \quad s_{1-2}\hat{S}_{12}^T] \, / \det j, \tag{eq. 2.82}$$

where $\hat{S}_{12}$, $\hat{S}_{23}$, and $\hat{S}_{31}$ are the coordinates of the lines \$$_{12}$, \$$_{23}$, and \$$_{31}$ passing through the points of intersection of the connector lines \$$_1$, \$$_2$, and \$$_3$, as illustrated in Figure 4.10. Now

$$(j^T)^{-1} = (j^{-1})^T) = [s_{3-2}\hat{S}_{23}^T \quad s_{1-3}\hat{S}_{31}^T \quad s_{2-1}\hat{S}_{12}^T]^T \, / \det j$$

$$= [s_{3-2}\hat{S}_{23} \quad s_{1-3}\hat{S}_{31} \quad s_{2-1}\hat{S}_{12}] \, / \det j$$

and thus from (4.36):

$$\hat{T} = [s_{3-2}\hat{S}_{23} \quad s_{1-3}\hat{S}_{31} \quad s_{2-1}\hat{S}_{12}] \begin{bmatrix} v_1 \\ v_2 \\ v_3 \end{bmatrix} \, / \det j. \tag{4.37}$$

We also show in Chapter 2 that $\det j$ can be expressed in three ways:

$$\det j = s_{3-2}\hat{S}_{23}^T\hat{s}_1, \tag{eq. 2.84}$$

$$\det j = s_{1-3}\hat{S}_{31}^T\hat{s}_2, \tag{eq. 2.85}$$

$$\det j = s_{2-1}\hat{S}_{12}^T\hat{s}_3. \tag{eq. 2.86}$$

The geometrical meaning of $\det j$ is now apparent in the sense that it can be expressed to a scalar multiple as any one of the reciprocal products of the pairs of lines ($\$_{23}$, $\$_1$), ($\$_{31}$, $\$_2$), and ($\$_{12}$, $\$_3$).

It is also useful to compute the component v_{ti} of the velocity of point C_i that is perpendicular to the line $\$_i$. The coordinates of the line $\$_{ic}$ are

$$\hat{s}_{iC} = \begin{bmatrix} -s_i \\ c_i \\ \ell_i \end{bmatrix}.$$

This line intersects the axis of the revolute joint at C_i, and it is also reciprocal to the sliding motion

$$\hat{T}_{2i} = v_i \begin{bmatrix} c_i \\ s_i \\ 0 \end{bmatrix}$$

because

$$\hat{s}_{iC}^T\,\hat{T}_{2i} = \hat{T}_{2i}^T\hat{s}_{iC} = v_i\,[-s_i,\,c_i;\,\ell_i]\begin{bmatrix} c_i \\ s_i \\ 0 \end{bmatrix} = 0.$$

Finally, the reciprocal product

$$\hat{s}_{iC}^T\,\hat{T}_{li} = \omega_{li}\,[-s_i\; c_i\; \ell_i]\begin{bmatrix} 0 \\ 0 \\ 1 \end{bmatrix} = \omega_{li}\ell_i = v_{ti}.$$

Hence, the formation of the reciprocal product of the left and right sides of

(4.30) with $\hat{s}_{ic}$ yields $v_{ti} = \hat{s}_{iC}^T \, \hat{T}$, and therefore the velocities v_{t1}, v_{t2}, and v_{t3} are given by

$$v_{t1} = \hat{s}_{1C}^T \, \hat{T},$$

$$v_{t2} = \hat{s}_{2C}^T \, \hat{T},$$

$$v_{t3} = \hat{s}_{3C}^T \, \hat{T}, \tag{4.38}$$

which can be expressed in the matrix form

$$\mathbf{v_t} = [C]^T \, \hat{T}, \tag{4.39}$$

where $\mathbf{v_t} = [v_{t1}, v_{t2}, v_{t3}]^T$ and $[C] = [\hat{s}_{1C} \; \hat{s}_{2C} \; \hat{s}_{3C}]$.

We may now deduce the corresponding results for an instantaneous rotation $\delta\phi$ of the moving platform about G. From (4.35):

$$\begin{bmatrix} \delta\ell_1 \\ \delta\ell_2 \\ \delta\ell_3 \end{bmatrix} = j^T \, \delta\hat{D}, \tag{4.40}$$

which is the expression for the vector of the displacements of the moving pivots C_i parallel to the lines $\$_i$. Clearly,

$$\mathbf{v} = \lim_{\delta t \to 0} \begin{bmatrix} \delta\ell_1/\delta t \\ \delta\ell_2/\delta t \\ \delta\ell_3/\delta t \end{bmatrix} \quad \text{and} \quad \hat{T} = \lim_{\delta t \to 0} \frac{\delta\hat{D}}{\delta t}.$$

From (4.39) the vector of the tangential displacements of the moving pivots C_i is given by

$$\begin{bmatrix} \ell_1 \, \delta\theta_1 \\ \ell_2 \, \delta\theta_2 \\ \ell_3 \, \delta\theta_3 \end{bmatrix} = [C]^T \delta\hat{D}. \tag{4.41}$$

Thus,

$$v_t = \lim_{\delta t \to 0} \begin{bmatrix} \ell_1 \, \delta\theta_1/\delta t \\ \ell_2 \, \delta\theta_2/\delta t \\ \ell_3 \, \delta\theta_3/\delta t \end{bmatrix}$$

It is instructive to deduce (4.40) and (4.41) from first principles, because

Table 4.1 *The dualities of parallel and serial manipulators*

Parallel	Serial
Wrench coordinates $\hat{w} = (\mathbf{f};\, \mathbf{c}_o)$	Twist coordinates $\hat{T} = (\mathbf{v}_o;\, \boldsymbol{\omega})$ $\delta\hat{D} = (\delta x_o,\, \delta y_o;\, \delta\phi)$
Twist coordinates $\hat{T} = (\mathbf{v}_o,\, \boldsymbol{\omega})$ $\delta\hat{D} = (\delta x_o,\, \delta y_o;\, \delta\phi)$	Wrench coordinates $\hat{w} = (\mathbf{f};\, \mathbf{c}_o)$
Scalar connector forces $\boldsymbol{\lambda} = (f_1, f_2, f_3)$	Scalar joint velocities $\boldsymbol{\gamma} = (\omega_1,\, \omega_2,\, \omega_3)$ Scalar joint displacements $\delta\boldsymbol{\theta} = (\delta\theta_1,\, \delta\theta_2,\, \delta\theta_3)$
Forward statics $\hat{w} = j\boldsymbol{\lambda}$	Forward kinematics $\hat{T} = J\boldsymbol{\gamma}$ $\delta\hat{D} = J\delta\boldsymbol{\theta}$
Connector velocities $\mathbf{v}$ Connector deflections $(\delta\ell_1,\, \delta\ell_2,\, \delta\ell_3)$	Joint torques $\boldsymbol{\tau}$
Inverse kinematics $\mathbf{v} = j^T\hat{T}$ $[\delta\ell_1,\, \delta\ell_2,\, \delta\ell_3]^T = j^T\delta\hat{D}$	Inverse statics $\boldsymbol{\tau} = J^T\hat{w}$

these results are important in the analysis of a compliant parallel manipulator (see Chapter 5). Before we proceed with this analysis, the instantaneous kinematics of parallel manipulators in singular configurations are analyzed and the dualities between serial and parallel manipulators are listed in Table 4.1.

4.6 Instantaneous kinematics of parallel manipulators in singularity configurations

We studied the statics of parallel manipulators in Section 2.9, which showed that a singularity condition occurs when the connector lines meet in a finite point Q. In addition, the three connector forces belong to a pencil of forces that pass through Q and are linearly dependent (see Fig. 4.14).

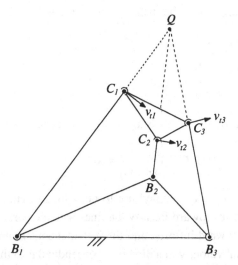

Figure 4.14 Singularity condition; the connector lines meet at Q.

In this configuration, the moving platform has an uncontrollable instant mobility and instantaneous angular velocity $\omega \neq O$ about Q. Because of this, points C_i ($i = 1, 2, 3$) have velocities $v_{ti} = \omega Q C_i$ perpendicular to the connector lines, and hence, the actuator sliding velocities v_i are zero.

This is easy to show by choosing Q as the origin of an xy coordinate system for which

$$ j = \begin{bmatrix} c_1 & c_2 & c_3 \\ s_1 & s_2 & s_3 \\ 0 & 0 & 0 \end{bmatrix}. $$

The substitution of this result into (4.35) together with

$$ \mathbf{v} = \begin{bmatrix} 0 \\ 0 \\ 0 \end{bmatrix} $$

yields

$$ \begin{bmatrix} 0 \\ 0 \\ 0 \end{bmatrix} = \begin{bmatrix} c_1 & s_1 & 0 \\ c_2 & s_2 & 0 \\ c_3 & s_3 & 0 \end{bmatrix} \omega \begin{bmatrix} y_G \\ -x_G \\ 1 \end{bmatrix}, \tag{4.42} $$

where (x_G, y_G) are the coordinates of the instant center. Expand (4.42) to get

$$0 = \omega \{y_G c_1 - x_G s_1 + 0\},$$

$$0 = \omega \{y_G c_2 - x_G s_2 + 0\},$$

$$0 = \omega \{y_G c_3 - x_G s_3 + 0\}. \tag{4.43}$$

For $\omega \neq 0$ the only solution of set (4.43) is $x_G = y_G = 0$ because generally

$$\begin{vmatrix} c_1 & s_1 \\ c_2 & c_2 \end{vmatrix} \neq 0, \quad \begin{vmatrix} c_1 & s_1 \\ c_3 & s_3 \end{vmatrix} \neq 0, \quad \begin{vmatrix} c_2 & c_3 \\ s_2 & s_3 \end{vmatrix} \neq 0,$$

i.e., any pair of connector lines are assumed not to be collinear. Hence, the instant center is Q. As Q moves toward infinity the lines become parallel and the instant center G moves toward infinity, and the instantaneous mobility is an instantaneous translational velocity in a direction perpendicular to the connector lines (see also Fig. 2.27 where this velocity is parallel to the y axis).

4.7   An infinitesimal displacement analysis for a parallel manipulator

Figure 4.15 illustrates the moving platform of a parallel manipulator undergoing an infinitesimal rotation $\delta\phi$ about an axis $ through a point G. This is modeled by a revolute joint at G (connected to the frame of reference in the fixed body) to which the moving platform is rigidly connected. The lines $\$_{iB}$ and $\$_{iC}$ are perpendicular to the lines of the connectors $\$_i$ ($i = 1, 2, 3$), and they pass through the fixed and moving pivots B_i and C_i, respectively.

When the moving platform rotates relative to the fixed platform about an axis through G, each moving pivot C_i displaces to a point C_i'. This displacement can be decomposed into two displacements, an infinitesimal displacement $\delta\ell_i$ along the line $\$_i$ together with an infinitesimal displacement $\ell_i\delta\theta_i$, which is tangent to a circle of radius ℓ_i centered at B_i (see Fig. 4.16). It should be clear from this figure that

$$\delta\ell_i = r_i\delta\phi \quad \text{and} \quad \ell_i\delta\theta_i = r_{iC}\delta\phi, \tag{4.44}$$

where r_i and r_{iC} are, respectively, the perpendicular distances from G to the lines $\$_i$ and $\$_{iC}$. Therefore

$$r_i = \hat{s}_i^T \hat{S}, \quad \text{and} \quad r_{iC} = \hat{s}_{iC}^T \hat{S}, \tag{4.45}$$

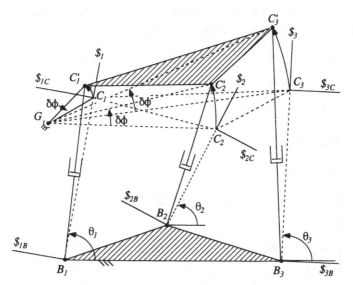

Figure 4.15 Infinitesimal displacement of a parallel manipulator.

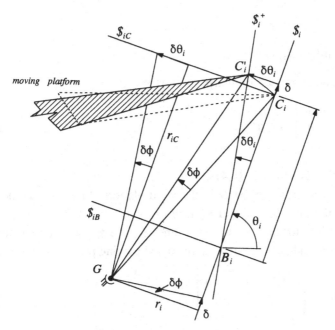

Figure 4.16 Infinitesimal displacement of a single connector.

where $\hat{s}_i$, $\hat{s}_{iC}$, and $\hat{S}$ are the coordinates of the lines $\$_i$, $\$_{iC}$, and the axis $\$$ through G. The substitution of (4.45) into (4.44) yields

$$\delta\ell_i = \hat{s}_i^T(\delta\phi\hat{S}) = \hat{s}_i^T\delta\hat{D} \quad \text{and} \quad \ell_i\delta\theta_i = \hat{s}_{iC}^T(\delta\phi\hat{S}) = \hat{s}_{iC}^T\delta\hat{D}. \qquad (4.46)$$

Hence, for $i = 1, 2, 3$,

$$\delta\ell_1 = \hat{s}_1^T\delta\hat{D},$$

$$\delta\ell_2 = \hat{s}_2^T\delta\hat{D},$$

$$\delta\ell_3 = \hat{s}_3^T\delta\hat{D}, \qquad (4.47)$$

which can be expressed in matrix form as

$$\begin{bmatrix} \delta\ell_1 \\ \delta\ell_2 \\ \delta\ell_3 \end{bmatrix} = j^T\delta\hat{D}. \qquad (4.48)$$

This equation is precisely (4.40). Also, for $i = 1, 2, 3$,

$$\ell_1\delta\theta_1 = \hat{s}_{1C}^T\delta\hat{D},$$

$$\ell_2\delta\theta_2 = \hat{s}_{2C}^T\delta\hat{D},$$

$$\ell_3\delta\theta_3 = \hat{s}_{3C}^T\delta\hat{D}, \qquad (4.49)$$

which can be expressed in matrix form as

$$\begin{bmatrix} \ell_1\delta\theta_1 \\ \ell_2\delta\theta_2 \\ \ell_3\delta\theta_3 \end{bmatrix} = [C]^T\delta\hat{D}. \qquad (4.50)$$

This equation is precisely (4.41).

Choose a reference point O in the fixed platform to determine the coordinates of the lines $\$_i$, $\$_{iB}$, and $\$_{iC}$ (see Fig. 4.17).

The matrices for the coordinates of lines $\$_{iB}$ and $\$_{iC}$, which are perpendicular to line $\$_i$ and pass through the fixed and moving pivots are, respectively,

$$[B] = \begin{bmatrix} -s_1 & -s_2 & -s_3 \\ c_1 & c_2 & c_3 \\ q_{1B} & q_{2B} & q_{3B} \end{bmatrix}, \qquad (4.51)$$

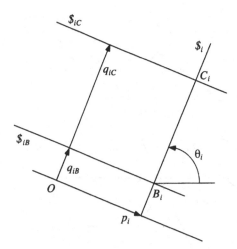

Figure 4.17 The location of lines $\$_i$, $\$_{ib}$, and $\$_{ic}$.

$$[C] = \begin{bmatrix} -s_1 & -s_2 & -s_3 \\ c_1 & c_2 & c_3 \\ q_{1C} & q_{2C} & q_{3C} \end{bmatrix},$$ (4.52)

where $q_{iC} = q_{iB} + \ell_i$.

Observe from Figures 4.15 and 4.16 that when the moving platform rotates through an angle $\delta\phi$, each connector moves from a line $\$_i$ to a new line $\$_i^+$ due to a small rotation $\delta\theta_i$ about the fixed pivot B_i. We can now determine the coordinates for $\$^+$.

4.8 The differential of a line

Assume that the line $\$_i$ drawn in Figure 4.18 is attached to a revolute joint located at point B_i and is connected to the ground (the page), which is the frame of reference. For a small rotation $\delta\theta_i$, line $\$_i$ moves to a second line $\$_i^+$ and every point on $\$_i$ undergoes a small rotation, except for the point that is coincident with B_i. Consider that the coordinates $\hat{s}_i$ of line $\$_i$ are parametric functions of θ_i, and

$$\hat{s}_i = \hat{s}_i\,(\theta_i).$$ (4.53)

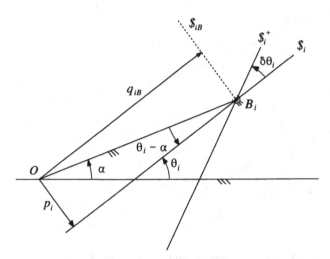

Figure 4.18 The differential of a line.

Therefore, the differential of the coordinates is

$$\delta \hat{s}_i = \frac{d\hat{s}_i}{d\theta_i} \, \delta\theta_i. \tag{4.54}$$

From Figure 4.18, the coordinates of $\$_i$ are

$$\hat{s}_i = \{c_i, s_i; p_i\} = \{c_i, s_i; OB \sin(\theta_i - \alpha)\}. \tag{4.55}$$

Write $\hat{s}_{iB} = d\hat{s}_i / d\theta_i$ to yield

$$\hat{s}_{iB} = \frac{d\hat{s}_i}{d\theta_i} = \{-s_i, c_i; OB \cos(\theta_i - \alpha)\} = \{-s_i, c_i; q_{iB}\}. \tag{4.56}$$

These are the coordinates for line $\$_{iB}$, *which is defined as the geometrical differential of* $\$_i$. In addition, the coordinates of $\$_i^+$ are given by

$$\hat{s}_i^+ = \hat{s}_i + \hat{s}_{iB}\delta\theta_i, \tag{4.57}$$

which are linear combinations of the coordinates of lines $\$_i$ and $\$_{iB}$.

5

The stiffness mapping for a parallel manipulator

5.1 A derivation of the stiffness mapping

Figure 5.1 illustrates an elastically compliant, planar parallel manipulator. The moving and fixed platforms are connected by three *RPR* serial chains and in each prismatic pair there is a linear spring. Assume that the moving platform is in equilibrium with an externally applied force with coordinates $\hat{w}$ and magnitude f is applied to it on a line $\$$. Then

$$\hat{w} = f_1\hat{s}_1 + f_2\hat{s}_2 + f_3\hat{s}_3, \qquad \text{(eq. 2.60)}$$

where f_1, f_2, and f_3 are the magnitudes of the resultant forces in the connectors and $\hat{s}_1$, $\hat{s}_2$, and $\hat{s}_3$ are the line coordinates of the connectors.

A small change $\delta\hat{w}$ in the applied force will cause the upper platform to move with an infinitesimal rotation with coordinates $\delta\hat{D}$ on an axis perpendicular to the page through a point G, as illustrated by Figure 5.2. These quantities are related by a 3×3 stiffness matrix $[K]$ which we will determine next.

Assume that the free lengths of the springs ℓ_{oi} and the stiffness constants k_i are known, together with the coordinates $\hat{w}$ of the applied force and the coordinates $\hat{s}_1$, $\hat{s}_2$, and $\hat{s}_3$ of the lines $\$_1$, $\$_2$, and $\$_3$. Assume also that the movable lamina is initially loaded and the spring lengths are ℓ_1, ℓ_2, and ℓ_3. Then substitute the relationships $f_i = k_i (\ell_i - \ell_{oi})$ $(i = 1, 2, 3)$ into (2.60), where k_i are the spring constants and $(\ell_i - \ell_{oi})$ is the difference between the current and free length of an ith spring, to yield

$$\hat{w} = k_1(\ell_1 - \ell_{01})\hat{s}_1 + k_2(\ell_2 - \ell_{02})\hat{s}_2 + k_3(\ell_3 - \ell_{03})\hat{s}_3. \qquad (5.1)$$

153

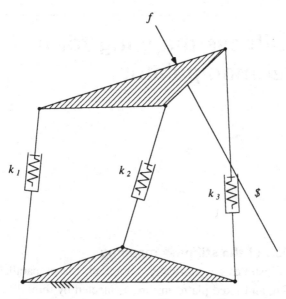

Figure 5.1 A planar compliant coupling.

The total derivative of (5.1) can be expressed in the form

$$
\begin{aligned}
\delta\hat{w} &= k_1\delta\ell_1\hat{s}_1 + k_2\delta\ell_2\hat{s}_2 + k_3\delta\ell_3\hat{s}_3 \\
&\quad + k_1(\ell_1 - \ell_{01})\frac{d\hat{s}_1}{d\theta_1}\delta\theta_1 + k_2(\ell_2 - \ell_{02})\frac{d\hat{s}_2}{d\theta_2}\delta\theta_2 \\
&\quad + k_3(\ell_3 - \ell_{03})\frac{d\hat{s}_3}{d\theta_3}\delta\theta_3 \\
&= \hat{s}_1 k_1 \delta\ell_1 + \hat{s}_2 k_2 \delta\ell_2 + \hat{s}_3 k_3 \delta\ell_3 + \hat{s}_{1B}k_1(1 - \rho_1)\ell_1\delta\theta_1 \\
&\quad + \hat{s}_{2B}k_2(1 - \rho_2)\ell_2\delta\theta_2 + \hat{s}_{3B}(1 - \rho_3)\ell_3\delta\theta_3,
\end{aligned} \tag{5.2}
$$

where $\rho_i = \ell_{oi}/\ell_i$ and $d\hat{s}_i/d\theta_i = \hat{s}_{iB}$ are the coordinates of a line $\$_{iB}$ perpendicular to $\$_i$ that passes through a fixed pivot B_i (see Fig. 5.2 and equation (4.56)). Equation 5.2 can now be expressed in matrix form as

$$
\delta\hat{w} = [\hat{s}_1\ \hat{s}_2\ \hat{s}_3]\ [k] \begin{bmatrix} \delta\ell_1 \\ \delta\ell_2 \\ \delta\ell_3 \end{bmatrix} + [\hat{s}_{1B}\ \hat{s}_{2B}\ \hat{s}_{3B}]\ [k(1 - \rho)] \begin{bmatrix} \ell_1\delta\theta_1 \\ \ell_2\delta\theta_2 \\ \ell_3\delta\theta_3 \end{bmatrix}, \tag{5.3}
$$

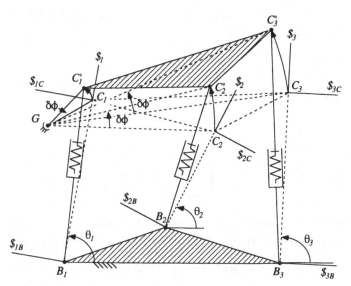

Figure 5.2 Motion of the platform with respect to a reference frame in the fixed platform.

where $[k]$ and $[k(1 - \rho)]$ are the (3×3) diagonal matrices

$$[k] = \begin{bmatrix} k_1 & 0 & 0 \\ 0 & k_2 & 0 \\ 0 & 0 & k_3 \end{bmatrix} \text{ and } [k(1 - \rho)] = \begin{bmatrix} k_1(1 - \rho_1) & 0 & 0 \\ 0 & k_2(1 - \rho_2) & 0 \\ 0 & 0 & k_3(1 - \rho_3) \end{bmatrix}.$$

The following notation was employed in Chapters 2 and 4:

$$j = [\hat{s}_1 \ \hat{s}_2 \ \hat{s}_3] = \begin{bmatrix} c_1 & c_2 & c_3 \\ s_1 & s_2 & s_3 \\ p_1 & p_2 & p_3 \end{bmatrix}$$

is the matrix of the coordinates of the lines $\$_1$, $\$_2$, and $\$_3$, and

$$[B] = [\hat{s}_{1B} \ \hat{s}_{2B} \ \hat{s}_{3B}] = \begin{bmatrix} -s_1 & -s_2 & -s_3 \\ c_1 & c_2 & c_3 \\ q_{1B} & q_{2B} & q_{3B} \end{bmatrix}$$

is the matrix of the coordinates of the lines $\$_{1B}$, $\$_{2B}$, and $\$_{3B}$ (see Fig. 5.2). Also, from (4.48) and (4.50):

$$
\begin{bmatrix} \delta\ell_1 \\ \delta\ell_2 \\ \delta\ell_3 \end{bmatrix} = j^T \delta\hat{D} \quad \text{and} \quad \begin{bmatrix} \ell_1\delta\theta_1 \\ \ell_2\delta\theta_2 \\ \ell_3\delta\theta_3 \end{bmatrix} = [C]^T \delta\hat{D},
$$

where

$$
[C] = [\hat{s}_{1C}\ \hat{s}_{2C}\ \hat{s}_{3C}] = \begin{bmatrix} -s_1 & -s_2 & -s_3 \\ c_1 & c_2 & c_3 \\ q_{1C} & q_{2C} & q_{3C} \end{bmatrix}
$$

is the matrix of the coordinates of the lines $\$_{1C}$, $\$_{2C}$, and $\$_{3C}$ (see Fig. 5.2). Substituting these expressions, (5.3) can be expressed in the abbreviated form

$$
\delta\hat{w} = \{j\,[k]\,j^T + [B]\,[k\,(1-\rho)]\,[C]^T\}\delta\hat{D}, \tag{5.4}
$$

and, finally, (5.4) can be expressed as

$$
\delta\hat{w} = [K]\,\delta\hat{D}, \tag{5.5}
$$

where the required stiffness matrix $[K]$ is given by

$$
[K] = j[k]j^T + [B]\,[k\,(1-\rho)][C]^T. \tag{5.6}
$$

Clearly, $[K]$ is symmetrical only at the unloaded position for which $\rho_i = 1$ or $\ell_i = \ell_{oi}$ and $[K] = j[k]j^T$.

It is easy to deduce from (3.33) that under the action of the Euclidean group, a twist $\delta\hat{D}''$ expressed in a new coordinate system is related to the same twist quantified in the original coordinate system by

$$
\delta\hat{D} = [E]\,\delta\hat{D}''. \tag{5.7}
$$

Similarly, an incremental change of force quantified in the new coordinate systems is related to the change of force quantified in the original coordinate system by (see (2.52))

$$
\delta\hat{w} = [e]\,\delta\hat{w}'', \tag{5.8}
$$

and

$$
j = [e]\,j''. \tag{5.9}
$$

Analogously, $[B] = [e] [B'']$ and $[C] = [e][C'']$. Substitute (5.7)–(5.9) into (5.4) to yield

$$[e] \, \delta\hat{w}'' = [e] \, \{j'' \, [k]j''^T + [B''] \, [k \, (1 - \rho)] \, [C'']^T\} \, [e]^T \, [E] \, \delta\hat{D}''. \qquad (5.10)$$

From (3.37), $[e]^T = [E]^{-1}$. Substituting this result in (5.10) and premultiplying the left and right sides by $[e]^{-1}$ yields

$$\delta\hat{w}'' = [K''] \, \delta\hat{D}'', \qquad\qquad\qquad (5.11)$$

where $[K''] = \{j'' \, [k] \, j''^T + [B''] \, [k \, (1 - \rho)] \, [C'']^T\}$. Equation 5.11 expresses the relationship between the twist produced by a change of force in a new coordinate system.

Equations 5.4 and 5.11 are expressions that relate the same twist expressed in terms of the old and new coordinate systems produced by the same force increment expressed in the same old and new coordinate systems. It is also clear that the stiffness matrix $[K]$ itself changes. Substitute (5.7) and (5.8) into (5.5) to give

$$[e] \, \delta\hat{w}'' = [K] \, [E] \, \delta\hat{D}'',$$

or

$$\delta\hat{w}'' = [e]^{-1} \, [K] \, [E] \, \delta\hat{D}'', \qquad\qquad (5.12)$$

and it follows from (3.37) that $[e]^{-1} = [E]^T$. Therefore, the new stiffness matrix $[K'']$ is related to the old matrix by

$$[K''] = [E]^T \, [K] \, [E]. \qquad\qquad\qquad (5.13)$$

5.2   The dimensions of the elements of the stiffness matrix

It is interesting to examine the dimensions of the elements of the stiffness matrix

$$[K] = j \, [k] \, j^T + [B] \, [k(1 - \rho)] \, [C]^T. \qquad\qquad (eq. \, 5.6)$$

The dimensions of the two terms on the right side of (5.6) are clearly identical because j, $[B]$, and $[C]$ are matrices whose columns are the coordinates of lines in the xy plane. Furthermore, ρ is dimensionless. Without loss of generality, the dimensions of the elements of the matrix $[K]$ will be determined for the symmetric case by substituting $\rho = 1$ in (5.6) and expanding the right side, which yields

$$[K] = \begin{bmatrix} c_1 & c_2 & c_3 \\ s_1 & s_2 & s_3 \\ p_1 & p_2 & p_3 \end{bmatrix} \begin{bmatrix} k_1 & 0 & 0 \\ 0 & k_2 & 0 \\ 0 & 0 & k_3 \end{bmatrix} \begin{bmatrix} c_1 & s_1 & p_1 \\ c_2 & s_2 & p_2 \\ c_3 & s_3 & p_3 \end{bmatrix}. \tag{5.14}$$

Therefore,

$$[K] = \begin{bmatrix} c_1 k_1 & c_2 k_2 & c_3 k_3 \\ s_1 k_1 & s_2 k_2 & s_3 k_3 \\ p_1 k_1 & p_2 k_2 & p_3 k_3 \end{bmatrix} \begin{bmatrix} c_1 & s_1 & p_1 \\ c_2 & s_2 & p_2 \\ c_3 & s_3 & p_3 \end{bmatrix}. \tag{5.15}$$

It follows that $[K]$ is a 3×3 symmetric matrix which can be expressed in the form

$$[K] = \begin{bmatrix} k_{11} & k_{21} & k_{31} \\ k_{21} & k_{22} & k_{32} \\ k_{31} & k_{32} & k_{33} \end{bmatrix}, \tag{5.16}$$

where expanding the right side of (5.15) and equating to the elements of (5.16) yields

$$k_{11} = \sum_1^3 c_i^2 k_i,$$

$$k_{21} = \sum_1^3 c_i s_i k_i, \quad k_{22} = \sum_1^3 s_i^2 k_i$$

$$k_{31} = \sum_1^3 c_i p_i k_i, \quad k_{32} = \sum_1^3 s_i p_i k_i, \quad k_{33} = \sum_1^3 p_i^2 k_i. \tag{5.17}$$

It is important to recognize that the dimensions (dim) of the elements of $[K]$ are not all the same: dim (k_i) = force/length = FL^{-1}, dim (p_i) = length = L, and dim $(c_i, s_i) = 1$. Hence

$$\dim [K] = \begin{bmatrix} FL^{-1} & FL^{-1} & F \\ FL^{-1} & FL^{-1} & F \\ F & F & FL \end{bmatrix}. \tag{5.18}$$

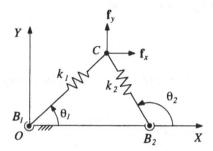

Figure 5.3 A two-spring system.

Now

$$\text{dim } \delta\hat{D} = \begin{bmatrix} L \\ L \\ 1 \end{bmatrix}.$$

(5.19)

Therefore, from (5.18) and (5.19):

$$\text{dim } ([K] \, \delta\hat{D}) = \begin{bmatrix} F \\ F \\ FL \end{bmatrix} = \text{dim } \delta\hat{w}.$$

(5.20)

The dimensions of (5.18) are thus consistent.

5.3   The stiffness mapping of a planar two-spring system

Figure 5.3 is a schematic representation of a pair of *RPR* serial connectors, with springs in the prismatic joints, and connections to the ground at points B_1 and B_2. The free ends are connected to a common turning joint at a point C.

This system can be obtained from Figure 5.1 by shrinking the movable platform to a point C and removing the third *RPR* connector. Equation 5.5 can then be expressed by

$$\delta f = [K] \, \delta D,$$

(5.21)

where

$$\delta f = \begin{bmatrix} \delta f_x \\ \delta f_y \end{bmatrix}$$

is a small increment of force acting at point C and

$$\delta \mathbf{D} = \begin{bmatrix} \delta x \\ \delta y \end{bmatrix}$$

is a small displacement of point C. The stiffness matrix can be expressed in the form

$$[K] = j [k] j^T + [C] [k(1 - \rho)] [C]^T, \qquad (5.22)$$

where all the submatrices are now 2×2, and

$$j = \begin{bmatrix} c_1 & c_2 \\ s_1 & s_2 \end{bmatrix}, \qquad (5.23)$$

$$[B] = [C] = \begin{bmatrix} -s_1 & -s_2 \\ c_1 & c_2 \end{bmatrix}. \qquad (5.24)$$

Furthermore,

$$[k] = \begin{bmatrix} k_1 & 0 \\ 0 & k_2 \end{bmatrix}, \qquad (5.25)$$

and

$$[k(1 - \rho)] = \begin{bmatrix} k_1(1 - \rho_1) & 0 \\ 0 & k_2(1 - \rho_2) \end{bmatrix}. \qquad (5.26)$$

It is left to the reader to deduce this result by writing the equation

$$\mathbf{f} = f_1 \mathbf{s}_1 + f_2 \mathbf{s}_2, \qquad (5.27)$$

where

$$\mathbf{f} = \begin{bmatrix} f_x \\ f_y \end{bmatrix} \quad \text{and} \quad \mathbf{s}_1 = \begin{bmatrix} c_1 \\ s_1 \end{bmatrix}, \quad \mathbf{s}_2 = \begin{bmatrix} c_2 \\ s_2 \end{bmatrix}.$$

Then introduce the spring constants k_1 and k_2 into (5.27), which yields

$$\mathbf{f} = k_1(\ell_1 - \ell_{01})\mathbf{s}_1 + k_2(\ell_2 - \ell_{02})\mathbf{s}_2. \qquad (5.28)$$

Let us take a total derivative of (5.28) (see (5.2)) and make the necessary substitutions by simplifying the results from Chapter 4.

Now assume that a wheel is connected to a platform using a two-spring

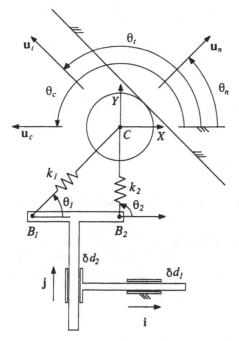

Figure 5.4 A passive two-parameter spring actuated by a P-P manipulator.

system, as illustrated in Figure 5.4. The platform is connected to the ground by a serial pair of actuated prismatic joints that are tuned for fine position control. The wheel maintains contact with a rigid wall.

From (5.6), $[K]$ is a symmetric 2×2 matrix and

$$[K] = [K_0] + [K_\rho],$$
(5.29)

where

$$[K_0] = \begin{bmatrix} c_1 & c_2 \\ s_1 & s_2 \end{bmatrix} \begin{bmatrix} k_1 & 0 \\ 0 & k_2 \end{bmatrix} \begin{bmatrix} c_1 & s_1 \\ c_2 & s_2 \end{bmatrix},$$
(5.30)

and

$$[K_\rho] = \begin{bmatrix} -s_1 & -s_2 \\ c_1 & c_2 \end{bmatrix} \begin{bmatrix} k_1(1 - \rho_1) & 0 \\ 0 & k_2(1 - \rho_2) \end{bmatrix} \begin{bmatrix} -s_1 & c_1 \\ -s_2 & c_2 \end{bmatrix},$$
(5.31)

where $\rho_i = \ell_{oi}/\ell_i$.

Now assume that the system is close to its unloaded position ($\ell_i = \ell_{oi}$). From (5.29):

$$[K] = \begin{bmatrix} c_1 & c_2 \\ s_1 & s_2 \end{bmatrix} \begin{bmatrix} k_1 & 0 \\ 0 & k_2 \end{bmatrix} \begin{bmatrix} c_1 & s_1 \\ c_2 & s_2 \end{bmatrix}. \tag{5.32}$$

Assume that the values $\theta_1 = 45$ degrees, $\theta_2 = 90$ degrees, $k_1 = k_2 = 10$ lbf/in. gives

$$[K] = \begin{bmatrix} 5 & 5 \\ 5 & 15 \end{bmatrix} \text{ lbf/in.}$$

Assume initially that the center point C is completely constrained by clamping the wheel to the wall. This means that any external force $\mathbf{f} = f_x \mathbf{i} + f_y \mathbf{j}$ acting through C can be applied to the wheel. Also, any small displacement of the lamina $B_1 B_2$, relative to the slider displacements δd_1 and δd_2, will cause a small change in $\mathbf{f}$ because of the changes in the lengths of the pair of springs. Now

$$\delta \mathbf{D} = \begin{bmatrix} \delta x \\ \delta y \end{bmatrix} = -\begin{bmatrix} \delta d_1 \\ \delta d_2 \end{bmatrix},$$

and therefore

$$\begin{bmatrix} \delta f_x \\ \delta f_y \end{bmatrix} = -\begin{bmatrix} 5 & 5 \\ 5 & 15 \end{bmatrix} \begin{bmatrix} \delta d_1 \\ \delta d_2 \end{bmatrix}. \tag{5.33}$$

Invert the matrix in (5.33) to give

$$\begin{bmatrix} \delta d_{c1} \\ \delta d_{c2} \end{bmatrix} = -\begin{bmatrix} 0.3 & -0.1 \\ -0.1 & 0.1 \end{bmatrix} \begin{bmatrix} \delta f_x \\ \delta f_y \end{bmatrix}. \tag{5.34}$$

The subscript c denotes correcting slider displacements which control the contact force.

Equation 5.34 can be used to control a time-varying contact force $\mathbf{f} = f_x \mathbf{i} + f_y \mathbf{j}$ between the fully constrained wheel and its environment. At each instant, an error in force is known which can be reduced using (5.34) by computing the proper force error-reducing displacement (or force-correcting displacement) $\delta \mathbf{D}_c = \delta d_{c1} \mathbf{i} + \delta d_{c2} \mathbf{j}$.

Assume now that the wheel is at rest, but that it is loaded with an excessive normal force parallel to the vector $\mathbf{u}_n$ (see Fig. 5.4). This force, which is reciprocal to the wheel motions, can thus be reduced without moving the wheel provided that

$$\delta \mathbf{f}_n = \begin{bmatrix} \delta f_x \\ \delta f_y \end{bmatrix} = \delta f_n \begin{bmatrix} \cos 45 \\ \sin 45 \end{bmatrix} = \delta f_n \begin{bmatrix} 0.707 \\ 0.707 \end{bmatrix}, \qquad (5.35)$$

where δf_n is the desired change in the normal force. Substitute this result in (5.34) to give

$$\delta \mathbf{D}_c = \begin{bmatrix} \delta d_{c1} \\ \delta d_{c2} \end{bmatrix} = \delta f_n \begin{bmatrix} -0.1414 \text{in./lbf} \\ 0 \end{bmatrix}. \qquad (5.36)$$

So, for this example, $\delta d_{c1} = -0.1414 \, df_n$ and $\delta d_{c2} = 0$. Therefore, a displacement in the negative x direction (for a positive δf_n) reduces a compressive normal force that passes through C. A compressive contact force f_n is negative and hence a positive δf_n reduces an excessively compressive contact force. It is important to recognize that the wheel does not move when the lamina $B_1 B_2$ moves in the x direction. *Hence, this is inherently the best direction for correcting a normal force error.* Intuitively, one may consider, at the outset, that a motion of the laminar parallel to $\mathbf{u}_n$ would change the contact force without moving the wheel. This is clearly not the case.

Finally, assume that the wheel C moves upward along the wall. The displacement of the wheel center C is given by

$$\delta \mathbf{D}_t = \delta d_{t1} \, \mathbf{i} + \delta d_{t2} \, \mathbf{j}, \qquad (5.37)$$

where $\delta d_{t1} = \delta p_t \cos 135$, $\delta d_{t2} = \delta p_t \sin 135$, and δp_t is a small displacement of point C parallel to the wall. Here the subscript t denotes the displacement $\delta \mathbf{D}_t$ of the wheel, which is tangential to the normal contact force produced by the slider displacements, δd_{t1} and δd_{t2}. Hence

$$\delta \mathbf{D}_t = \delta p_t \, (-707\mathbf{i} + 0.707\mathbf{j}), \qquad (5.38)$$

and expressing (5.38) in matrix form yields

$$\delta \mathbf{D}_t = \delta p_t \begin{bmatrix} -0.707 \\ 0.707 \end{bmatrix}. \qquad (5.39)$$

It is important to recognize that $\delta \mathbf{D}_t$ is the only allowable freedom for point C, assuming that contact between the wheel and the wall is to be maintained.

Also, a change in the constraint force $\delta \mathbf{f}_n$ cannot produce motion. As stated earlier, such forces and displacements are reciprocal and

$$\delta \mathbf{f}_n^T \delta \mathbf{D}_t = \delta f \, [0.707,\ 0.707] \, \delta p_t \begin{bmatrix} -0.707 \\ 0.707 \end{bmatrix} = 0.$$

It follows that the small slider displacements $\delta \mathbf{d}_c$ and $\delta \mathbf{d}_t$ can be used for the simultaneous control of normal force and tangential motion. By super-position

$$\begin{bmatrix} \delta d_1 \\ \delta d_2 \end{bmatrix} = G_c \, \delta f_n \begin{bmatrix} -0.1414 \\ 0.0000 \end{bmatrix} + G_t \, \delta p_t \begin{bmatrix} -0.707 \\ 0.707 \end{bmatrix}. \tag{5.40}$$

In (5.40) G_c and G_t are dimensionless scalar gains, and δp_t and δf_n are errors in the wheel position and the normal contact force.

5.4 Force and motion control using a serial manipulator with a compliant wrist

Figure 5.5 illustrates a planar three-revolute serial manipulator with a compliant wrist. The workpiece, which is held fixed in the gripper, is in contact with a fixed rigid lamina at a single point P. Hence, there is a single constraint force that acts on the workpiece along the line $\$_a$ and the workpiece thus has two freedoms: a pure translational displacement along the surface of the lamina and a rotation about the contact point P. Assume that the workpiece is to remain in contact with the rigid lamina.

The compliant wrist is an in-parallel manipulator with special geometry (see Fig. 5.6). Coaxial turning joints B_1 and B_2 are labeled B_{12} and the coaxial turning joints C_2, C_3 are labeled C_{23}. The forward analysis for this device is given in subsection 1.6.3. Each of the three *RPR* connectors has linear springs.

The revolute joints of the serial manipulator are actuated, and changes in the angular displacements produced by the actuators are denoted by ($\delta \psi_1$, $\delta \psi_2$, $\delta \psi_3$). Assume the manipulator is tuned for fine position control so that it is nonback-drivable when serving. The three *RPR* connectors are not actuated in the application, and the wrist is thus passive. The relative location of the pair of platforms is controlled by actuating the joints of the serial manipulator.

Assume initially that the tool is fully constrained by clamping it to the

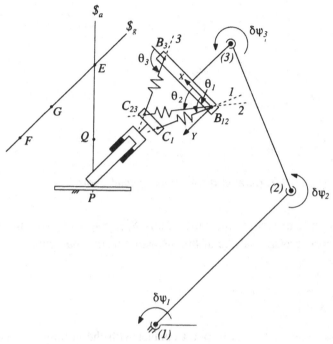

Figure 5.5 A passive three-parameter spring actuated by a 3R manipulator.

ground. This means that any force with coordinates $\hat{w}$ applied to the work-piece by the ground does not move the platform or robot manipulator. However, any twist of the movable lamina ($B_{12} - B_3$) with coordinates $\delta\hat{D}_F$ measured relative to the ground will change the force applied to the workpiece, and this can be expressed by

$$\delta\hat{w} = -[K]\delta\hat{D}_F. \tag{5.41}$$

The negative sign was introduced because the twist of the platform ($C_1 - C_{23}$) relative to the grounded workpiece was specified previously. Assume that the contact force between the workpiece and the ground is to be reduced by an amount $\delta\hat{w}$. The coordinates of the required twist $\delta\hat{D}_F$ can be computed from (5.41) and

$$\delta\hat{D}_F = -[K]^{-1}\,\delta\hat{w}. \tag{5.42}$$

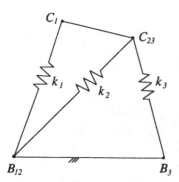

Figure 5.6 A parallel device with special geometry.

Following this, the required joint motions $\delta\psi_1$, $\delta\psi_2$, and $\delta\psi_3$ can be computed from a reverse analysis of the 3R manipulator, and using (3.52):

$$\begin{bmatrix} \delta\psi_1 \\ \delta\psi_2 \\ \delta\psi_3 \end{bmatrix} = J^{-1}\,\delta\hat{D}_F. \tag{5.43}$$

Suppose that the tool is in point contact with the ground at P (see Fig. 5.5). The line of action of the normal constraint force is labeled $\$_a$. We need to determine a point F about which the base movable lamina will rotate to control (and even reduce to zero, if required) the constraint or contact force at P. From (5.42):

$$\delta\hat{D}_F = -[K]^{-1}\,\delta\hat{w}_a, \tag{5.44}$$

where $\delta\hat{w}_a = \delta f[c_a,\ s_a;\ r_a]^T$ and $\delta\hat{D}_F = \delta\phi[y_F,\ -x_F;\ 1]^T$. The required point of rotation F with coordinates (x_F, y_F) is determined from (5.44) by the stiffness mapping $[K]$.

A rotation of the movable lamina about F cannot move the tool because the change of force applied to it is acting on the line $\$_a$. *Point F is therefore the best point to use to control the contact force because a twist with coordinates* $\delta\hat{D}_F$ *cannot cause motion of the tool.*

Assume finally that the base movable lamina undergoes an infinitesimal twist about an axis through some point G. This twist can be decomposed into a twist of the workpiece about an axis through point E, together with a twist

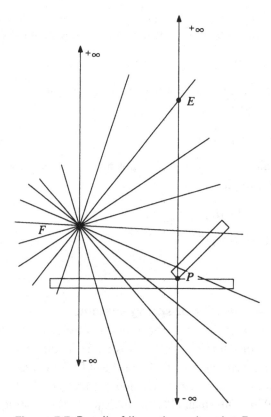

Figure 5.7 Pencil of lines through point *F*.

about an axis through point F (see Fig. 5.5) which simply alters the contact force along $\$_a$, and

$$\delta \hat{D}_G = \delta \hat{D}_F + \delta \hat{D}_E. \tag{5.45}$$

The line $\$_g$ that joins points F and G intersects $\$_a$ at point E. The workpiece twists an amount $\delta \hat{D}_E$ $(= \delta \hat{D}_G - \delta \hat{D}_F)$ relative to the ground.

Point G can lie anywhere in the plane. This means that a pencil of lines can be drawn through point F, and there are corresponding points of intersections E on the line $\$_a$ in the range $-\infty$ to $+\infty$ (see Fig. 5.7).

If a point G is selected, such that the line FG is parallel to S_a, then point E lies at infinity and the motion of the workpiece is a pure displacement along the ground for which $\delta \hat{D}_E = \delta \mathbf{D}_E = \{\delta x,\ \delta y;\ 0\}$. Point E must lie on the line $\$_a$. This is the condition of reciprocity for which the instant work must vanish:

$$\delta \hat{D}_E^T\ \delta \hat{w}_a = \delta \phi \delta f_a\ (y_E c_a - x_E\ s_a + r_a) = 0. \tag{5.46}$$

It follows from (5.45) that the law for the simultaneous control of motion and force can be expressed as

$$\delta \hat{D}_G = G_E \delta \hat{D}_E + G_F \delta \hat{D}_F, \tag{5.47}$$

where G_F and G_E are dimensionless gains for position and force errors. Substitute (5.44) into (5.47) to yield

$$\delta \hat{D}_G = G_E\ \delta \hat{D}_E - G_F[K]^{-1}\delta \hat{w}_a$$

$$= G_E \delta \hat{D}_E - G_F \delta f[K]^{-1}\hat{s}_a, \tag{5.48}$$

where $\delta \hat{w}_a = \delta f \hat{s}_a$.

5.5   A note on the stability of spring systems

A study of the instability of spring systems is beyond the scope of this text. However, it is of interest to note that some instability phenomena of the two-spring system were reported by Pigoski and Duffy (1993). Earlier relevant work on stability was reported by Haringx (1942, 1947, 1949, 1950), Eijk and Dijksman (1976), Dijksman (1979), and Eijk (1985). Stability phenomena that result in sudden erratic behavior can be explained by a mathematical theory called catastrophe theory (see Zeeman 1977, Arnol'd 1992, and Bruce and Giblin 1993). Catastrophe theory is currently being applied to the two-spring system by R. Hines (a research assistant at the University of Florida), D. Marsh (a lecturer in mathematics at Napier University, Edinburgh), and the author.

As far as the author is aware there is no study of the stability of the planar-three-spring system in progress. A relevant paper on the planar three-spring system was reported by Griffis and Duffy (1992).

───── **EXERCISE 5.1** ─────────────────────────

The two-spring system (see Exr. Fig. 5.1) is in its unloaded configuration. A vertical force F is applied at C. Plot the locus of the equilibrium posi-

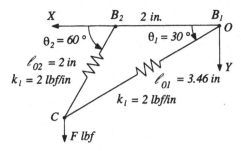

Exercise Figure 5.1

tions for C when the magnitude of the force increases in increments of 0.25 lbf in the range $0 \le F \le 5$ lbf.

(a) Assume that the spring matrix is given by

$$[K_0] = \begin{bmatrix} c_1 & c_2 \\ s_1 & s_2 \end{bmatrix} \begin{bmatrix} k_1 & 0 \\ 0 & k_2 \end{bmatrix} \begin{bmatrix} c_1 & s_1 \\ c_2 & s_2 \end{bmatrix}.$$

(b) Assume that the spring matrix is given by

$$[K] = [K_0] + \begin{bmatrix} -s_1 & -s_2 \\ c_1 & c_2 \end{bmatrix} \begin{bmatrix} k_1(1 - p_1) & 0 \\ 0 & k_2(1 - p_2) \end{bmatrix} \begin{bmatrix} -s_1 & c_1 \\ -s_2 & c_2 \end{bmatrix}.$$

References

Arnol'd, V. I. 1992. *Catastrophe Theory*. New York: Springer-Verlag.

Ball, R. S. 1900. *A Treatise on the Theory of Screws*. Cambridge: Cambridge University Press.

Bruce, J. W., and P. J. Giblin. 1993. *Curves and Singularities*. Cambridge: Cambridge University Press.

Chung, Y. S., M. Griffis, and J. Duffy. 1994. "Repeatable Joint Displacement Generation for Redundant Robotic Systems." *Trans. ASME Journal of Mechanical Design* **116**, no. (1), (March):11–16.

Dijksman, J. 1979. "A Study of Some Aspects of the Mechanical Behavior of Cross-Spring Pivots and Plate Spring Mechanisms with Negative Stiffness." Ph.D. Dissertation, Delft University of Technology, Amsterdam, The Netherlands.

Dimentberg, F. M. 1968. *The Screw Calculus and Its Applications in Mechanics*. U.S. Department of Commerce FTD-HT-23-1632-67.

Eijk, J. 1985. "On the Design of Plate Spring Mechanisms." Ph.D. Dissertation, Delft University of Technology, Amsterdam, The Netherlands.

Eijk, J., and J. Dijksman. 1976. "Plate Spring Mechanisms with Constant Negative Stiffness." *Vakgroep Fijnmechanische Techniek*, June.

Griffis, M., and J. Duffy. 1992. "Comparing Structures of Stiffness Matrices Using Invariants." Presented at the VIII CISM-IFToMM Symposium "Ro. Man. Sy '92."

Haringx, J. 1942. "On the Buckling and the Lateral Rigidity of Helical Compression Springs." *Proc. Ned. Akad. Wetenschappen*, Amsterdam **45**.

Haringx, J. 1947. "On Highly Compressible Helical Spring and Rubber Bars and Their Application in Vibration Isolation." Ph.D. Dissertation, Delft University of Technology, Amsterdam, The Netherlands.

Haringx, J. 1949. "Elastic Stability of Helical Springs at a Compression Larger than Original Length." *Applied Scientific Research* **A1**, 417–434.

Haringx. J. 1950. "Instability of Springs." *Phillips Technical Review*, **11**(8).

Hunt, K. H. 1978 1990. *Kinematic Geometry of Mechanisms.* Oxford: Clarendon Press, 2–3, and 108–110.

Klein, F. 1939. "Elementary Mathematics from an Advanced Standpoint. Geometry," 25–66, and 86–88. New York: Macmillan.

Li, S., and G. K. Matthew. 1987. "A Kinematic and Dynamic Investigation of the Planar Assure II Group." In *Proceedings of the Seventh World Congress on the Theory of Machines and Mechanisms*, E. Bautista, J. Garia-Loman, and A. Navarro (Eds.). Oxford: Pergamon Press, 141–145.

Pigoski, T., and J. Duffy. 1993. "An Inverse Force Analysis of a Planar Two-Spring System." Presented at the First Austrian IFToMM Symposium, Seggauberg, Austria (4–9 July). Published 1995. *Trans. ASME Journal of Mechanical Design* **117** (December).

Reuleaux, F. 1876. *The Kinematics of Machinery*, p. 47. New York: Macmillan.

Zeeman, E. 1977. Catastrophe Theory: *Selected Papers, 1972–1977.* Boston: Addison-Wesley Advanced Book Program.

Index